ENVIRONMENTAL ENCOUNTER

**Experiences
in decision-making
for the
built and
the natural
environment**

**Joanne Henderson Pratt
James Pratt
Sarah Barnett Moore
William T. Moore, M.D.**

1982 Printing.

Copyright © 1979 by Reverchon Press, P.O. Box 19647, Dallas, Texas 75219. 214—526-6609. All rights reserved. Except for pages specifically noted as permissable for reproduction, no part of this book may be reproduced, stored in a retrieval system, or transcribed, in any form or by any means — electronic, mechanical, photocopying, recording, or otherwise — without the prior written permission of the publisher. Pages 88, 89, 115, 146 and 147 may be reproduced for use as questionnaires for field trips in connection with the use of this book.

ISBN 0-9601902-0-1
LC Catalog Card No. 78-62125

Text typeface: *11/12 Garamond Bold #3*
Margin Typeface: 10/10 Garamond Displays: *Garamond Ultra.* Printed and bound in the United States of America.

"The Chair," and "The Ceiling," copyright 1950 by Theodore Roethke from THE COLLECTED POEMS OF THEODORE ROETHKE. Used by permission of Doubleday & Company, Inc. The excerpt from "Twelve Precepts of Modern Design" is used by permission, Edgar Kaufmann, Jr., WHAT IS MODERN DESIGN? The Museum of Modern Art, New York, Copyright © 1950, renewed 1977 by The Museum of Modern Art, New York. "Postscript" to Prologue: The Birth of Architecture," Copyright © 1965 by W. H. Auden. Reprinted from ABOUT THE HOUSE, by W. H. Auden, by permission of Random House, Inc. "The Spite Fence" is used by permission of Richard Eberhart. The quotation from the interview with Elizabeth White is used by permission of Maggie Wilson, copyright 1974. The excerpt from THE ART OF COLOR by Johannes Itten is used with permission of the Reinhold Publishing Company, copyright 1961. The passage from THE STRANGE LAST VOYAGE OF DONALD CROWHURST is used with permission of Stein and Day, publishers, copyright 1970.

acknowledgements

We wish to express our gratitude to the National Endowment for the Arts for the grant that has made possible the publication of this book, and to the University of Texas at Arlington under whose auspices three grants have been administered to help develop *Environmental Encounter*.

To the United States Office of Education for two grants, under the Environmental Education Act of 1970, to test materials with faculty from twelve departments, and students at the University of Texas at Arlington.

To the American Institute of Architects, for its grant that supported initial testing of materials and a first manuscript; to the Dallas Chapter of the American Institute of Architects that funded a three-year collaboration between school teachers and architects; to the Perkins School of Theology of Southern Methodist University, the Dallas Independent School District, and The Greenhill School whose faculty and students participated in formulating the learning strategies.

To David Connally, who proposed the first graphic concept for layout of the book and contributed time and ideas that remain a part of the design as developed and executed by James Pratt. To Rex Hendershot and Beverly Swan for graphic advice and encouragement. To Bob Taylor whose cartoons enliven the book and to Ray Goodrow whose drawings help say what the photographs cannot.

To Deans Charles Green, Harold Box, and George Wright and their faculties who hosted the authors' work in their schools at the University of Texas at Arlington, as well as Dr. James White and Jo Faye Godby of Southern Methodist University, and Evelyn Beard of the Dallas Independent School District.

To those many individuals who have been supportive including David Clarke, former Executive Director, Association of Collegiate Schools of Architecture, Alan G. Levy, Dr. Aase Eriksen, Milton C. Powell, Rose Marie Banks, Judy Glazer, Edmund N. Bacon, Stanley Madeja, Rosemary Schroeder, James Ellison, Bill Sheveland, George Pearl, Jack Luby, Mary Ann Rumbo, Richard Ferrier, Laura Devlin, Downing Thomas, Bill, Jean and Betsy Booziotis, Quinten Mathews, Mel Armand, Mary John Barnett, Ilya Pratt, Jerry and Philip Henderson, and Thomas Taylor; to the staff at the Dallas Public Library who were unflagging in their enthusiastic help, Linda Robinson, Lois Hudgins, Frances Bell, and Judith McPheron.

To the members of the National Committee on Public Education of the American Institute of Architects who over a four-year span were a sounding board to help focus a direction for this work, and to members of the Dallas Chapter of the American Institute of Architects who contributed their time to workshops for primary and secondary school teachers.

To those who helped put the text in its final form: Sally Wiley whose generous time devoted to editing the copy sharpened our words and clarified our meanings, and Oriealice Strait and Carol Naab for their typing and support.

Table of contents

Overview

Environmental Encounter is designed to reveal the unseen relationship between people and their environment. Humans perceive what they have a predisposition to perceive. They do not often notice, for example, how they are affected by the qualities of the spaces in which they live — their color, texture, shape, and scale. This book expands the individual's awareness of often ignored aspects of the interaction with one's surroundings. The book emphasizes the physical elements of this relationship, while acknowledging the impact of the social, the psychological, and the physical on one another. As people become more aware of these elements, they will learn to evaluate and make more informed decisions about themselves and their universe.

Environmental Encounter begins by asking us as individuals participating in a series of planned experiences to be aware of ourselves as sensory beings in relation to environment. The focus of the exercises then widens to include ourselves in a community of individuals. Finally, the exercises help us examine our relationship with environment in its broadest sense.

Each chapter has a theme with several corollary concepts underlying the learning strategies. These themes are presented in the "Conceptual Strands." In succeeding chapters, concepts appear and reappear, permeating and tieing together the learning experiences. Any one of these conceptual strands, or perhaps others, could be used as the basis for sequencing the learning experiences. The authors have chosen the following sequence:

orientation

Chapters One and Two introduce the concept that each of us has more resources in our past experience than we realize. These resources can contribute to our understanding of environment. The exercises invite each of us to analyze why particular physical spaces have had special meaning for us at different periods in our lives.

Participants then choose a quality, such as tranquillity or security, build a space suggesting that quality, and together analyze the result.

receiving and organizing sensory data

We make contact with the external world through our sensory receptors. Chapter Three explores sensory stimulation, a process we undergo continuously, but which we are only occasionally consciously aware of. To avoid an overload of sensations, we unconsciously order these sensory "receptions" into "perceptions." For example, although we are receiving thousands of stimuli at any given moment, we are conscious only of a generalized state of being. "Well-being" might include "touching" a seventy-three degree atmosphere, "seeing" a well-lighted newspaper, "tasting" a clean mouth, "smelling" coffee brewing, and "hearing" familiar household voices.

Chapters Four through Eleven ask us to analyze perceptions to demonstrate what techniques we are actually using, consciously or unconsciously, to order sensory stimuli and thus save ourselves from sensory chaos.

Chapters Twelve through Fifteen examine ordered wholes: an array of small objects, larger built environments, and nature. The learning experiences help sort out subjective impressions ("I like it," "I don't like it") and past associations ("It reminds me of . . ."). The experiences then lead us to analyze how the parts of a composition come together to form a whole. Chapter Sixteen asks teams of participants to apply the concepts previously learned to build an environment of contrasting spaces. Each team's space is designed to evoke a particular sensory response and yet relate to the whole Sequence of Spaces.

living in community

Our operational, psychological, and social needs cause us to seek a community in which to live. Chapters Seventeen through Twenty-five examine the impact of neighborhood and the larger urban environment on our lives and our impact on them. In building environments, choices must be made that in our society often bring into conflict public and private values. Being both "private" individual and "public" citizen, each of us must evaluate any choice from both points of view.

synthesis

Prior learning strategies in the exercises have isolated and exaggerated experiences in the environment so they could be perceived and analyzed. Chapter Twenty-six explores the nature of change and the implications of living in our constantly changing environment. Chapter Twenty-seven asks us to expand perceptions of what constitutes our world and to take personal responsibility for what we do with it.

This final synthesis is a summing up; but it also can be a beginning, because we spend our entire lives increasing the breadth of our experience and sharpening our tools for living in environment.

Conceptual strands
underlying the learning strategies

Environmental decisions must be based on a balancing of many factors. In the past, economic and other special interests have tended to narrow environmental decision-making. Although plant and animal preservation are increasingly considered, attention must be given also to the impact of the full range of environmental decisions on the quality of our lives, and on our physical and psychological well-being.

We have few tools for measuring and few theories for predicting the consequences on our lives of our environmental decisions. There are, however, some valid generalizations. These are brought up specifically at various points in the book; most of them underlie almost every decision we make.

verbal communication about environment is fragmental

Verbal communication, being a linear process in time and limited to the symbolism of words, constricts the ability to express a total experience.

visual communication about environment is fragmental

It is a two-dimensional communication about a four-dimensional subject. The third dimension and time both make it impossible to completely portray an environment in a photograph or drawing. One must be taught the conventions of portrayal in order to grasp two-dimensional drawings. The many non-visual stimuli are missing.

to establish order, we must understand and communicate our own need for order

We seek order for psychological and physical comfort. We constantly strive, consciously or otherwise, to relieve anxiety by comprehending our personal relationship with at least our immediate environment. At home we sink contentedly into "our" chair.

our only contact with our environment is through our sense receptors

These are our visual, aural, tactile, olfactory, gustatory, kinesthetic, and proprioceptive senses. Automatically, stimuli, *received* by our receptors are *perceived* in patterns. Sound waves *received* by our ears as discrete high frequencies may be *perceived* as a scream.

perception is reception plus association

For a United States citizen, to perceive the "Star Spangled Banner" is to hear this music and stand up. To a citizen of Java, the music may be perceived as Western, but he will have no association motivating him to rise. A Westerner may receive the frequencies that make up Javanese music but perceive its patterns only as being foreign, with no other association.

discovering patterns is
a way to find order

Seeking patterns consciously is an act of orienting in the environment. We look for the street pattern as we arrive in a city. If it is radial, instead of the familiar grid, it takes us much longer to get our bearings.

we are our own yardstick

We function on a relative basis, with ourselves as the reference point. We constantly compare new experiences with past experiences, new objects with objects perceived in the past. We look for the fit that feels just right.

we are a moving reference point

Even when we assume we are at rest, our eyes continuously survey objects through peripheral vision. We see with "eyes in the back of the head." The heart always beats. We are always in motion as we perceive environment.

our reliance on a particular hierarchy
of the sense receptors is culturally
derived and becomes habitual

The highly differentiated cultures of the twentieth century depend far more on the eye than older or less complex cultures. Speed helps to create this dependence on the eye and atrophies the use of the senses of smell, touch, and taste. We could hear, smell, and touch the environment from a trotting buggy. The experience is entirely different from a speeding car.

we perceive by similarity
and by contrast

Rhythm is the repetition of similar units within a pattern. It allows us to easily understand many parts by inference from one part that we clearly understand already, without attending every one in detail. A dissimilar unit of a pattern stands out by contrast, forcing us to notice it. In a building facade with repeating identical windows, the door is emphasized by a contrast in size and placement so that it is easily found.

we perceive all objects
within a context

Personal satisfaction derived from our perception of a whole is particularly dependent on the interrelationship of its parts. If the parts do not act together to create an integrated whole, we are left with an uneasy feeling. A plastic vibrator-recliner chair with all its movable parts is disturbing in the context of a French Provincial living room.

How many of us can "see" the neighborhood in the context of the city? The problem is not only one of design. The neighborhood and the city as a whole are interdependent. Ultimately the vitality of one depends on the vitality of the other. Because we are trained to look only at the objects, not the context, we ignore the hints of deterioration and so are shocked by the full-scale blight. The failure to see a building within the context of the city puts a strain on the economic and social health of the city as a whole.

in trying to manipulate simultaneously all the parts — both objects and processes — we lose our perception of the whole

This perceptual limitation makes it difficult to intelligently guide the growth or change of the complex system, our total environment. The urban designer attempts to develop physical order out of the many physical and social systems that make up a city. The ecologist must manipulate innumerable variables in the framework of the whole biosphere.

our reaction to design is often influenced by our unconscious associations

An ex-soldier may abhor the color khaki. We may dislike a particular tea cup because it reminds us of childhood and having to clean our plates down to the flowers. A connoisseur of tea cups, however, sees beyond his personal associations to appreciate the form, texture, quality of glaze, color, method of manufacture, economy of line, and decoration. The ability to perceive and to differentiate many components within a whole reflects a high level of sophistication in the viewer.

we tend simultaneously to synthesize and to analyze any experience

Seeing a room for the first time we perceive a total impression. Simultaneously, we analyze its components: the chair fabric, ceiling height, pattern of windows and floor. Again, simultaneously, we may strongly respond to a particular component of the room and mentally transfer it to our own remembered environment. "Wouldn't that table be nice by my bed?"

we use selective inattention to help us adapt to discordance and to excess information in our environment

Our senses receive great quantities of information that our perceptive processes do not allow to come to consciousness. We do not see the clutter of strip development. If we live beside a freeway we do not hear the traffic that gives a houseguest a sleepless night.

we are a part of our environment

We are as one with our environment and our environment is as one with us. We consume oxygen from the atmosphere and exhale carbon dioxide to it. Even after death, our bodies continue in an exchange with the environment.

each of us has a personal perception of our common environment

As individuals our perceptions differ from those of others. Our experiences in the environment are unique to each of us. A child looks at snow and sees a hill to sled down; his father perceives only a driveway to be cleaned. What we eat, the climate, our jobs, the conditions we work in, the people around us — all influence how we perceive and act in our environment. "We shape our buildings; thereafter they shape us."

orientation is defined here as the ability to perceive oneself in relation to the patterns of environment

Orientation is one aspect of our need for order. Physical orientation relates the self to time and space. The environment can hinder or facilitate orientation. A forest or a new subdivision usually gives us little help because of the sameness of the trees or the houses. A clock tower, on the other hand, signals our distance from a fixed place visually and aurally. The clock orients us in time by a commonly adopted pattern: four bongs for four o'clock. The tower, by its age, adds a historical orientation.

our ability to make independent decisions differentiates us from other natural species

We can predict the probable consequences of our action (science) and make a judgment about the effect of our act (ethics and morality). The felling of a tree represents a decision a beaver does not have to make.

as individuals we require an area of personal space around us wherever we are

Our personal space (our space bubble) surrounds us as we move; it expands or contracts according to our needs at a particular time and in a particular situation. We step away from an enemy; we embrace a close friend. If there is not enough space to support our psychological needs in a particular social context, we become stressed. We can't sit still.

People of different cultures have different personal-space needs. Italians, for example, approach each other very closely, gesturing both love and anger at an intimate distance.

Our personal space bubble grows as we age. The very young are without strong images of self, and have yet to develop a need for a defined space zone.

territorial space, as distinguished from personal space, implies a relationship to a particular physical area

Space may be viewed as territory that over a period of time belongs to the individual or is shared — sometimes both. "Who's parked his car in *my* space?" This relationship ties us emotionally to the physical place we value.

design is an ordered relationship of components

Design is a way we organize sensory data into meaningful wholes. When we line a road evenly with trees, we make a functional and esthetic design. We also create a design when we draw the road and the trees.

A design will be deeply satisfying only if the beholder can respond from the perspective of his own culture to the direct qualities of the form of the design. The tea ceremony in a meticulously designed room is ultimately satisfying to the Japanese who has trained his body and mind to the discipline of following a rigidly prescribed pattern. To a Westerner it can be an incomprehensible ceremony that he appreciates only because it is exotic.

order reduces stress

We orient ourselves by comprehending the patterns of order of the spaces we are in. This reduces our feeling of stress as we encounter new environments. Too much order, however, eventually leads to boredom and we demand new stimuli. The accumulation of new stimuli leads us again to disorder.

We choose styles that swing from modern starkness to Victorian clutter and back. In such a dynamic process a sense of balance is achieved only momentarily. In our buildings as well as in our clothing, fashions are in constant transition, giving us the reassurance of perceivable order but the stimulation of new forms.

quality is in the eye of the beholder

Quality is relative; it entails a value judgment. For a particular person it changes as his perspective changes. With exposure to a broader range of possibilities, he finds that many of the objects chosen in his youth come to have only sentimental value.

The individual's assumptions of quality exist in a social context at a given moment; that is, they exist as an implicit standard defined by a particular culture or subculture. The implicit standard will vary with the degree of differentiation of the culture. It will also vary with time.

The level of environmental quality is defined here as the degree to which the environmental components interrelate with the whole. A continuous grade-separated network of bicycle paths is of higher quality than a system that requires conflict with automobile traffic.

agent, medium, and process are always part of a product, whether each is explicit or not

We are most aware of agent, medium, and process when we ourselves make a product, such as a clay pot. Those elements are not so apparent if the "product" is a built environment, such as a shopping center. Careful integration of elements leads to a product that functions successfully, whether to adorn a mantlepiece or to make shopping a pleasant experience.

In the making of an environment, the product is *explicit* although the agents, media, and process may not be at all apparent to the user. If they are apparent, however, experiencing the product is so much the richer.

In viewing artifacts of past cultures, such as the pyramids, the fascination lies, in part, in trying to discern how they came into being. How was the stone quarried, moved, and set in place? Similarly, we are intrigued with present building. Who is not at heart a sidewalk superintendent? Who does not peep through the batterboards at a building site and vicariously help to build the environment?

Agent, medium, process, and product each has importance to the user. When each is considered, there is greater likelihood that the man-made product can be experienced as an integrated whole, that it will work within the context of its environment, and that it will continue to function well for a long time.

symbols are shortcuts
to essential information

Symbols carry information and associated meanings. We are all conscious of clothes as symbols; the "grey flannel suit" or "long hair and patched jeans" clearly connote in our minds the persons within the clothes.

We are less conscious that the environment also carries symbolic messages. It can have obvious messages such as high walls with locked gates, which strongly imply "keep out." It can also convey more subtle messages; a fountain, for example, "invites" you to refresh yourself.

It is because the environment is symbolic to us that design can be used as a medium for sending messages. A restaurant achieves rapid customer turn-over with uncomfortable seats. Darkness and padded seating invite lingering.

nature is a source for design

Nature is our inspiration for design. This applies to both small-scale objects and large-scale environments. It is also the context in which we view man-made forms. San Francisco is satisfying because of the integration of the built with the natural surroundings. A caucasian rug abstracts natural forms and colors, and is at home in the Persian desert.

The idea that man can control nature is an illusion. Nature persists while man's efforts are finite. Man can bulldoze the streambed to build, but he cannot stop nature from inexorably washing a new path over the years.

Nature challenges us to enhance or maintain natural processes rather than interfere with them. Nature has a high degree of order, but has an infinite number of variables. Nature also undergoes constant hourly, daily, and seasonal change throughout time. It is difficult to predict the impact of our man-made building, but we should use all the tools at our disposal to try. Our reward is a life in harmony with nature, which provides the infinitely varied sensory stimuli that are so satisfying to us.

entertainment of the mind
is part of human experience
and is enhanced or inhibited
by our environment

Before the industrial revolution many natural and geometric symbols were incorporated into our built environments. Egyptian bas-reliefs and Roman mosaics aided our ability to maintain interest in built environments. They continually refreshed the eye, and stimulated the imagination. The emphasis in architecture for the past fifty years has been on new kinds of structure, building processes, and materials. We have come to depend directly on nature for visual stimulation with see-through and mirrored glass walls and indoor gardens.

constancy is change

We are in a dynamic interaction with our always changing environment. The weather changes from wet to dry; buildings are torn down and new ones built. We continuously seek to perceive the size, form, and changes occurring in the structure of our physical and social environment. We do this to maintain our orientation. Extreme rates of change, whether fast or slow, block our ability to perceive and thus make us uncomfortable.

Preface

Our focus for this book is the individual and how he or she relates with environmental systems. Central to any comprehensive understanding and ability to make informed decisions, must be our awareness that we are in a relationship with an environment that is not static; we are part of a dynamic process.

Our concern is for that part of our environment in which we spend ninety percent of our time and in which we are the principal agent — the built environment. Our awakening concern for ecology is encouraging. Still, a more all-encompassing awareness is missing. The relationship of the built environment to nature's systems, and even more important, our relationship to both, must be seen as parts of a whole.

This book is organized to help bring into focus those aspects of environment that normally lie just outside the periphery of our consciousness — but that we encounter daily. The book is conceived as a set of experiences in the environment to be undertaken and then analyzed.

Your particular concern will automatically determine how you use this book. The basis for each learning experience is an activity. How the learning experience is handled and the direction of discussions and analysis will come from you.

The activities and discussion in each learning experience are for those seeking to expand their environmental awareness. You may use this book for that purpose in a variety of ways. A professor may expand the twenty-seven chapters into a course using his expertise and emphasis as an integral part of what takes place. A facilitator may conduct a short workshop. Any person can use the book for self-instruction. You can browse through the pictures and text or act as instructor-participant by carrying out the learning experiences and asking yourself the questions relating to the experience. You become the "group" that the "instructor" teaches.

Whichever role you choose you bring to it your own knowledge and experience. They form your frame of reference and become the basis of an expanding, more all-encompassing response to the environment. That is our hope.

Spatial history

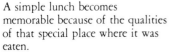

Understanding that environment affects us: recalling places from our past and present experience demonstrates the impact of spaces upon the activities that occur within them.

remember that place where?

Specific places where significant acts have occurred can have special meaning for us. There are times when the place where something important happens interacts with the event to reinforce the experience. The event would be different if it occurred in a different place. The interaction between ourselves and a particular place produces a lasting impression that affects our responses to similar places encountered later in our lives. Over the years each of us builds up his own spatial history.

A conscious awareness of the affect special places have upon us helps us to understand how all places affect us.

A simple lunch becomes memorable because of the qualities of that special place where it was eaten.

favorite spaces

Ask the group to rearrange the room to create an environment most conducive to discussion and to sharing personal experiences. Help them as little as possible, perhaps by leaving the room for ten minutes while they work out their solution. On returning, ask them why they think this is the best possible arrangement.

Ask the participants to close their eyes and picture in their minds the responses to the following questions. Give them time between each question to reflect.

"Think about the most pleasant place you frequent. Think about the shape of that place, its colors, its textures, the lighting and its effect, the odors, the temperatures, how the space is divided or defined, how you move through it, and what happens to you while you go through it. What kinds of things happen there? Does the place facilitate these happenings or distract from them?

"Think about the place that meant the most to you when you were a child, say six to eight years old." (Ask the same set of questions about that place.) Repeat, selecting ages appropriate to the group; for example, thirteen, nineteen, forty-five.

Ask each participant to select his favorite of those places he has recalled. Ask him to describe it to the group and discuss what it has meant to him.

What makes a favorite place?

our experience

The participants have great difficulty deciding as a group how to change the room.

The way you rejoin the group after they have rearranged the room sets the tone for the whole course. If a "teacher's place" had been arranged by the group, we chose not to use it. Instead, we moved into the area designated for the participants.

If the arrangement of the room was stiff and functioned poorly for our purposes, we probed for comments from the participants. The comments often resulted in moving, perhaps to the floor, in more relaxed positions.

"Which of the many qualities of that place made it special and why? Are there any places where you go today that have some of those same qualities? How do you feel about those places? How often do you try to 'recreate' that special feeling in a space today? Aside from the nostalgia involved, are those 'today' places really special in their own right? What are their positive and negative qualities?"

recalling spaces built as a child

Ask the participants to sketch personal spaces they built as children (forts, tree houses, caves . . .).

Point out that although it is difficult to represent space in two dimensions, it is even more difficult to communicate the qualities of light, texture, odor, and so forth, that helped make the place special. Ask each participant to explain his drawing to the group.

"What, as a child, were you trying to create?" (a refuge, a clubhouse . . .)
Did you succeed? Why or why not?

"At what point do you stop calling a place 'natural' and call it 'man-made' or 'built'? We say a virgin forest is natural, but what about a planted forest? a forest with campsites? a tree house?"

our experience
The important point to stress in this discussion is how the physical qualities facilitated interaction among the participants.

Do you recall the joy of creating your own small private place?

preparation
See notes for educators, page 164.
The meeting space should have
movable furniture.

further exploration

"Identify your personal space at work, at home, or at school.

"Explore 'What is the environment?' with a polaroid camera." Discuss the participants' photographic definitions of the environment with the entire group. To use this exercise as a pretest, compare each photographic presentation with the participant's results from the similar assignment in Photointerpretation, Chapter 14.

check list

Did the participant join the group to rearrange the room?

Was the individual able to express the feelings evoked in him by a particular place or physical environment?

Did the individual recognize to some extent how important to any human activity is the place where it is experienced?

resources

Bacon, Edmund: *Design of Cities*, Penguin Books, New York (1976). Richly illustrated definitive discussion of the evolution of great urban form ideas. The entire volume is rewarding reading. Specific pages will be cited as resources. Pages 15-19 cover awareness of space as experience.

Capote, Truman: *Other Voices, Other Rooms*, A Signet Book, New American Library, New York (1960). Description of personal spaces.

Evans, G. and Howard, R.: "Personal Space," *Psychological Bulletin* (October 1973) 334-344. Review article with extensive bibliography.

Proust, Marcel: "Overture" to *Swann's Way, In Remembrance of Things Past*, Vol. 1, Random House, New York (1934). Description of sensuous memories.

Sissman, L. E.: "Going Home," *Dying, an Introduction*, Little, Brown, Boston (1967). A poem describing returning home to the poet's childhood room.

Twain, Mark (Samuel L. Clemens): *The Adventures of Tom Sawyer*. The use of personal spaces by children.

Wolfe, Thomas: Chapter 1, *Look Homeward Angel*. Description of train ride home.

Creating personal space

2

Becoming aware of our ability to change our environment: construction of cardboard spaces designed to evoke a particular response is preparation for beginning to judge and value all spaces.

"we shape our buildings; thereafter they shape us."

As Winston Churchill expressed it, we are a product of the environment we build around ourselves. As we move into a space we make it ours for the moments or years we occupy it. But at the same time the space we occupy is affecting us; for example, making us feel comfortable and relaxed, or formal and constrained. We create a personal space out of even the most anonymous motel room. We throw our clothes on the bed and make the air fragrant with our favorite scent. We have the capacity to create personal spaces with qualities that will enhance our lives.

Neiman-Marcus specialty store Christmas catalog, 1973. "Give the impossible Dream: PRIVACY

This most sought-after and elusive of all commodities becomes a reality with a private world — designed to your needs and specifications . . . . Within your world, you relax in an atmosphere psychologically fitted to help you accomplish peak mental and personal potential . . . ."

The artist's private world on the right would represent an unrelaxing clutter to an accountant. Your personal space must fit *you*.

"we shape our buildings…"

Until we have defined space, we cannot perceive it. We define it, or shape it, by enclosing or demarking it in some way. We may not make "visible" the space, but only its edges, or objects in and around it. We build to enclose space, creating an "inside" that separates us from the "outside."

defining space

The idea in enclosing space is to separate ourselves from the out-of-doors.

defining space

In this pediatricians' clinic, glass separates the professional staff from noisy patients and the out-of-doors, but leaves them in visual contact with both. How many spaces can the person sitting at the typewriter look into?

Enclosed space is three-dimensional. It has shape established by the walls, ceiling, and floor, which defines a volume. We also build many partial enclosures for our use. A deep wing chair defines an adult minispace volume with only an implied completion of the space. A four-poster bed, a telephone or restaurant booth, a contemporary office work station each defines a particular volume of space. It may be finite and closed, or it may overlap with neighboring spaces and only imply a complete volume.

a street is an exterior room

A street can be more than a two-dimensional corridor for cars. It can be consciously designed for a unified spatial effect. A street lined with trees or buildings has the qualities of an enclosed space. It can be perceived as a volume or exterior room.

defining space
On the left tall buildings form the walls of a curving space.
On the right In late afternoon, cars are banned from this street while the Perugians come out to see and be seen in their city living room.
Below rhythmic tall trees planted along the entrance to a French chateau suggest movement toward infinity.

People in motion are themselves components that further define a space. Activities that take place in the street — children playing, trucks thundering, trees blossoming or shedding leaves — are as much a part of the whole as the buildings and light posts. Thus, the space changes with time, acquiring an additional physical quality, a patina of use. A patina will also build up of symbolic associations based on good or bad experiences in the space. The assassination of President Kennedy continues to affect the user's perception of Dealey Plaza in Dallas.

"...thereafter they shape us."

Spaces impart qualities to influence our feelings and actions: for example, intimacy, grandness, vastness, and claustrophobia can be suggested by space. One important quality space can impart is a sense of security. The need for security is part of us. We reveal our psychological need for it by the way we position ourselves in space. We sit frontier saloon style with our backs to the wall, the position of maximum security. In sleep, our most defenseless state, we lie with our heads to the wall. The need is emotional as well as functional.

Each individual's first secure space is the womb. Early man's cave was also a place of security; that is, shelter from the elements, from animals, and from other men. Part of this aspect of security is the feeling of privacy, of being hidden. The sense of security is heightened if "you" can see "them," but "they" cannot see "you."

quality of space
An automobile gives us a small-scale enclosure with a feeling of partial privacy, a sense of isolation in a comfortable sealed bubble, and a relative sense of control of space. These combine to give a sense of security even as our vehicle hurtles through space.

Our impetus to manipulate space is based on more than our seeking security. We need stimulation for our senses on the one hand, and on the other we need orderliness or at least our own perception of orderliness, in the things that stimulate us so that we are thrown neither into monotony nor into chaos. The small, dark spaces that interested us repeatedly as children would quickly bore us as adults. Such spaces lack the variety of sensory stimulation to which we are accustomed. Yet we often become bored in time with even the richest space. It no longer intrigues us; we have perceived its order and demand for ourselves new stimuli. We require variety not only in what we see, but also in what we hear, smell, taste, and touch, and in the way spaces cause us to move in and through them. Hence we need many kinds of spaces around us and a changing environment.

every space conveys silent messages

A built environment carries a nonverbal message; for example, "be welcome," "stay out of here," "be humble here," or "here be afraid." The message is twofold: It *may* be a conscious message from the maker to the beholder; it is *always* an unconscious message about the maker, his capabilities, and what he values. All such elements as color, texture, and patterning help to convey the particular message desired. A message of "security," for example, is more than a heavy bolt on the door. It may involve the thickness of doors and walls, the firmness of the floor, the sense of structural solidity of walls and roof, the size of openings into the space, the degree of privacy, and many more elements. Shops that wish to appeal to a particular group of people choose their entrances, display fixtures, and colors to convey the right message.

message of space
As you glimpse this view into a courtyard from a street in Rome what messages do you read?

The nonverbal message may be as direct as a locked iron fence expressing "stay out!" But a white picket fence overgrown with wisteria may convey the opposite message. If the picket fence triggers associations with a friend's welcoming home, it will be seen as an invitation to enter. Public spaces also communicate these messages, both directly and by association with our past experience. A library can demand formal demeanor with lined-up chairs and tables. Or it can encourage browsing with comfortable chairs grouped around area rugs.

Manipulating space is a satisfying activity. It is a challenge to do well. How we do it, either as individuals working for ourselves or as members of a group working for society, affects us all.

25

manipulating space

Begin with a brief discussion of (1) how space affects us and (2) how we can manipulate space to achieve desired qualities. Referring back to the spatial histories and to the text, help the participants define what constitutes a small personal space.

"What are the physical characteristics of a personal space that make it so satisfying? When you subtract the colors, the odors, the texture, what are the qualities of the *space* itself?

"Each space has entrances and exits, and exists in a context. How does the way you move into and through the space affect your mood? How does the positioning of the space in its context affect the impact within the space? When you look out of the space, what do you see? Does light from the windows enhance the mood of the space or harm it?

"In what ways could you manipulate only cardboard to 'shape spaces' in order to define a volume with a particular impact? For example, what feeling does a person have occupying a space with a very high ceiling? with a very low ceiling? with a skylight? with light coming only from slits in the walls? with a conical roof? with a turret? How does each alteration change the affect the space has on a person within it?

"To understand space as having characteristics that are independent of its enclosing boundaries, try thinking of it not as a void but as a solid form."

our experience

We found that this learning experience is a natural place for the participants to see that their personal taste is not necessarily a solution to a posed problem. Their personal preferences are valuable ipso facto; they are theirs. However, because they are so highly personal they do not necessarily communicate with other people.

Working individually with each participant two or three times during this exercise helped him realize his idea for enclosure. Some participants need help in simplifying a too elaborate construction, others in pushing their ideas further.
"What do you need to do now to make your space feel more _____?
"Don't be afraid to change, experiment, or start over.
"It's a big problem — you've a right to feel frustrated."

This learning experience is valuable for insights on how to use space efficiently. Working out the logistics of storing cardboard until the next session is of equal value. Spaces may be so constructed with duct tape that they can be disconnected at one or two junctures, folded, and stacked in a corner between uses.

creating a space which evokes a specific quality

Ask each participant to build a three-dimensional structure large enough to contain himself using cardboard and duct tape.

"Built environments need to be related to the individuals using them. In this project you must decide what particular qualities you desire in your personal space. Evoke an exaggerated response such as 'discomfort,' 'claustrophobia,' or 'extreme serenity.' How could you make someone feel _____? Keep your quality secret until finished."

Demonstrate various techniques of manipulating cardboard — taping, stapling, folding, and making simple cut joints.

When the projects are finished, exchange personal spaces, leaving in each a card bearing the adjective best expressing the emotions felt by the visitor. After the participants have compared their visitors' responses with their own intended effects, help them in a discussion to evaluate how successful they were in communicating their ideas. Now ask the participants to discuss larger scale environments in the same terms. "Identify rooms, buildings, and open spaces you have experienced that successfully evoked a sense of personal space."

Visit rooms and spaces immediately available. Analyze the "messages" the spaces communicate. "Are the messages appropriate to the function of these places?"

creating a village

Ask the group to create a "village" with their spaces, paying special attention to the whole composition. Stress that the structures be positioned to best enhance the qualities of each as well as the relationships among all of them. Emphasize that the group should not be satisfied until they find a solution they can verbally justify. They should pay particular attention to the space defined by the exteriors of the personal spaces, its shape, and how well it functions for the uses to which it will be put.

Discuss the village in terms of the spaces formed and its whole composition. Apply the same analysis to a familiar shopping complex and, finally, to the downtown center of the city.

leading questions

"Did you successfully evoke for others the emotional response you intended? For example, did your visitors have the same idea of comfort as you did?"

"Did you create the effect you set out to create? How well does it work?"

"How did the materials you worked with affect what you could do? If you were to rebuild it, would you do anything differently?"

"What aspects of design did you encounter?" (shape, volume, light, heat . . .)

"Did the odor of the cardboard or tape affect your response?"

"How did the working space affect what you could do?"

"How would this have been different if you had been working in a room by yourself?" (So many "favorite" spaces are places of solitude.)

"What is the difference between the outside and the inside of your structure? Did anyone create a piece of sculpture (form perceived only from the outside) rather than a space?"

preparation

double-faced corrugated cardboard — three 5 feet by 7 feet sheets per participant or team of participants

duct tape (2 inch wide plastic-coated adhesive)

staple guns and staples (optional)

metal or wood straight edges

matte knives

white butcher paper ("poster board") for sketching or for using as a material

The space in which this session is conducted should be large enough so that each participant has an area of at least 8 feet by 8 feet in which to work.

Save several enclosed and roofed spaces, created by the participants, to use in Chapter 8, Light.

Cardboard is easily cut and folded in several different ways that increase its strength. Warped planes, bent edges, and folded surfaces all result in stronger construction than the simple flat sheet.

warped plane

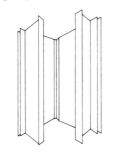

bent or ribbed edges

ribbed or folded surfaces

further exploration

Architecture students can draw a plan, perspective, and section of their space.

Engineering students can determine the structural limits of cardboard as a material. "How could your structure be made to carry more weight? Which parts are in tension, in compression, subject to shear or moment forces?"

English students can describe in prose or poetry how the built structures achieved the qualities felt within the spaces.

Psychology students can compare behavior in the personal spaces.

Teachers can explore uses of cardboard in their classrooms.

check list

Was the participant able to achieve the desired quality of personal space as evaluated (a) by himself, and (b) by his peers?

Has the participant used design components such as texture, light and color, pattern, movement through space, and shape of space to create the desired qualities, or has he used only objects to symbolize those qualities?

Was the participant able to analyze ways in which physical components interact to form environments with specific qualities?

resources

Bacon, Edmund: *Design of Cities*, Penguin Books, New York (1976). See Pages 15-19, 40-41, and 46-47 on the psychology of space.

Experiencing the Environment, S. Wapner, S. B. Cohen, and B. Kaplan (editors), Plenum Press, New York (1976). Research reports that reveal ways psychologists measure interactions between people and environment.

Farallones Scrapbook, Farallones Designs, Star Route, Point Reyes Station, Calif. 94956. Distributed by Random House (1971). Suggests constructions from many materials, and explains how to bend, tape, slot, and lace cardboard.

Hartung, Rolf: *Creating With Corrugated Paper*, Reinhold, New York (1966). Ideas for textured sculpture and objects made with cardboard.

Human Behavior and Environment, I. Altman and J. F. Wohlwill (editors), Plenum Press, New York (1976). Advances in theory and research.

Leonard, Michael: "Humanizing Space," *Progressive Architecture* (April 1969) 128-133. Discussion and accompanying diagrams cover sensory perception, movement, and sequence of spaces.

Norberg-Schulz, Christian: *Existence, Space, and Architecture*, Praeger, New York (1971). A development of the idea that architectural space may be understood as a concretization of environmental images which form a necessary part of man's orientation in the world. See Pages 9-12 for a historical development of concepts of man in space and Pages 30-31 and 86-89.

Psychology and the Built Environment, D. Canter and T. Lee (editors), Architectural Press, London (1974).

Sommer, Robert: *Personal Space*, Prentice Hall, Englewood Cliffs, N.J. (1969). The behavioral basis of design.

Sensory perception

3

Receiving and interpreting information: exercises that expand our understanding of how we use our sense receptors reinforce a conscious recognition that we are continuously stimulated by the environment.

only through our senses...

There is no other way to obtain information. Our senses receive stimuli from the environment and send impulses to the brain. The needle pricks our finger; we feel pain.

The *tactile, olfactory,* and *gustatory* receptors register relatively close stimuli. The *visual* and *auditory* receptors can, in addition register more distant stimuli. Two internal stimuli are usually taken completely for granted until for some reason they are subjected to unusual forces or are lost through disease: The *proprioceptors* are situated inside the body in tissues. They give us our sense of position; for example, whether our fingers are straight or bent, or whether we are standing or lying. The *kinesthetic* receptors account for the sensation of movement.

Perception is the interpretation of sensory impulses through past experience. The unseen prick registered by us as a syringe needle in our arm, might be assumed by a primitive tribesman as having come from a poisoned dart.

In a technological culture, visual perception is commonly used to interpret phenomena that have previously been confirmed directly by the other senses. This is called *visual interpretation.* The bread "looks" hard. The candy "looks" sweet. The mountain "looks" far away.

We use perception to orient us in time and space. Hearing a carillon tune, we know that it is noon and we are near the downtown. We base this conclusion on our past experience of having learned the location of the bell tower and the time the carillon is played.

distant and close receptors

Introduce the concepts on the preceding page by beginning a discussion of sensory perception.

"To become more aware of how we use our senses try listing all of them. Which are useful at relatively close distances? Which can be used when both long and short distances are involved? Are there any ways except through these senses that we can receive information?

"We perceive through direct stimulation of our sense receptors. We also get information by imagining the impact on one sense receptor by stimuli received by another. We see (visual sense) a fur coat and know it 'feels' (tactile sense) soft, without needing actually to touch it. By using our past experience we have made a visual interpretation of a tactile sensation. Can you give examples of visual interpretation of other senses?

"Every item is perceived in a context. Each 'object' exists in its field. In fact, its context is one of the major tools we use for deciding what we have perceived. The larger the context, the harder it is to perceive. In large contexts, for example, we may notice the house, but not its landscaping, or how it is related to its surroundings.

"What are some examples of situations in which the 'object' remains the same, but its field is different and thus the whole, including the object, is perceived differently?" (A fashionable woman would be perceived as high society if seen at the opera; as a hooker, if seen in a red-light district.)

description of partner

Pair off the participants, stationing the members of each pair two hundred to three hundred feet apart. Ask each person to record every detail he can perceive about the other.

"Walk one hundred feet towards each other. Repeat the description of your perception. Walk to within ten feet of each other and repeat the description.

"Walk still closer together. You should now be near enough to your partner that he is visually out of focus. Close your eyes and leave them closed. First one partner, then the other, describe your perceptions, concentrating as totally as possible on only your auditory sense. Repeat, using the olfactory sense, then the tactile sense. Open your eyes. Describe what you perceive visually. Move apart about one foot and note any perceptual differences from your previous positions."

Now you, the instructor, repeat the preceding exercise with a participant. Identify all the elements you can perceive, including the environment around the individual: shadows, form, scale, odor, clothing patterns . . . .

"Notice I'm describing the object and also the field in which it exists." As you move closer, describe your perception of your partner's movement: his gait, movement of arms and legs, and the sound of movement.

Point out to the participants that at the closer distance of ten feet, perception of form, color, scale, and texture seems to change considerably. Sound has become more noticeable. Also note for the group that visual perception of movement may cause one to "hear" skirts swishing. Visual perception of a "shiny" skirt may suggest that its material is silk or nylon.

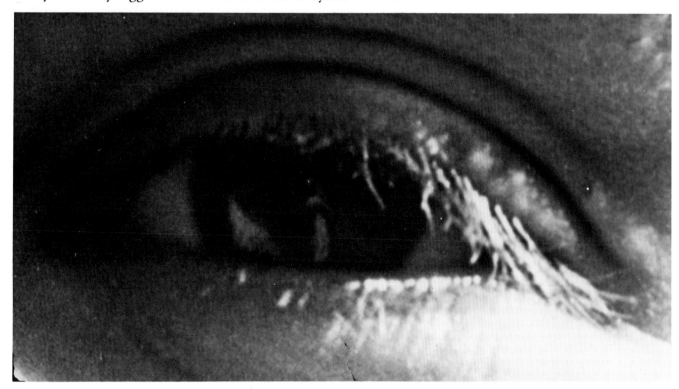

Comment on your use of various perceptions at zero distance: auditory (sounds of breathing, of cloth movement); olfactory (perfume, body odor, fabric odor, tobacco); tactile (texture of clothes, skin, firmness of fabric); gustatory (interpretation of peppermint flavor from odor of gum); proprioceptive (body balance, sense of position); and kinesthetic (forward movement).

Recapitulate for the group examples of how the direct perception of texture, sound, and movement differs from the visual interpretation of each. "What happens to each when the context changes?" (Direct perception is more reliable than visual interpretation, particularly if the context is unexpected. A plant indoors might be "seen" as plastic. The same plastic plant outdoors probably would be "seen" as real. In both contexts, the plant would be directly perceived by touch as being made of plastic.

our experience

The participants might paint their faces to exaggerate the effects in "description of partner."

We found it necessary to constantly emphasize the difference between reception, perception, and visual interpretation.

Except for the "blind walk," these exercises are best performed indoors because the distracting stimuli outdoors make concentration difficult.

blind walk

Ask participants to remove their shoes and socks and pair off. Ask one partner to act as the "seeing eye" who must both challenge and protect his blindfolded companion as he leads him on a walk. The chosen route should provide a variety of sensory stimulation, (for example, movement: elevator, stairs, ramps; texture: grass, concrete, tile; sound: birds, cars, voices). The blindfolded participant should identify objects along the route to his partner for later discussion. The partners should trade roles halfway through the allotted time period. At the end of the activity, discuss the sensory discoveries. Focus the participants' attention on the dominant role of the eye over other senses.

recognition of an object

Ask each participant to make a list of words describing a common object with unexpected properties; for example, an unusual saltshaker. Itemize in one column the descriptive words that result from direct sensory experience. Place each word under the sensory receptor used to obtain it, such as "visual: shiny, transparent; tactile: smooth, hard." In a separate column, write associative words such as "plastic," "antique."

"Which column contains the most adjectives? Which contains the least? Why? What does this imply as to the relative importance of our different sense receptors? Look at a texture and then touch it. Is it different from what you expected?"

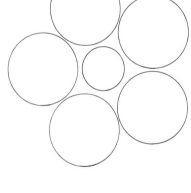

fooling the eye

"Play a trick on one of your senses. We are most familiar in our society with the trompe l'oeil because we are so visually oriented. Ask the participants to gather felt-tipped pens, a straightedge, and paper. "Draw several lines of equal length. Now add lines so as to make the original lines appear to be unequal. Have you succeeded in distorting the length of the lines as you see them at first glance? Experience tells you that the lines are still of equal length. You are fooled because from one sensory experience you are getting two different messages.

"Draw two circles of equal size, some distance apart. Surround the first circle by five larger equal circles and the second by five smaller, equal circles. Do the original circles still appear of equal size? Again your eye is fooled. Draw other geometrical shapes and try to distort them with superimposed lines. Try also to create an illusion of depth. You cannot draw a long corridor straight as it actually is. You must trick yourself. Did you achieve your illusion of distance with parallel lines or did you find it necessary to draw converging lines? The illusion of three dimensions is created on a two-dimensional sheet of paper by drawing lines not parallel as they are in reality, but converging at 'vanishing points.' Since the Renaissance, architects have created intentional optical illusions to make space seem larger.

"Now try to create the illusion of equivocal figures. Draw a checkerboard of repeating units of one shape. Color the shapes alternately black and white. Stare at the pattern you have drawn. Does it shift back and forth between a black pattern on a white ground and a white pattern on a black ground?"

statues

"Remember the game you played as a child called 'statues'? Someone would take you by the arm and throw you into an out-of-balance position that you would have to hold? Do this to a partner or assume an unbalanced position.

"What sensations do you feel? In which direction do you feel you will fall? Upon which joints is there stress? What is the feeling at your elbows, shoulders, hips, knees, ankles? Where do you wish to move next from this position?"

To achieve a temple that looks "straight" and has more apparent height, the Greeks curved the columns inward toward the top and bowed the architrave and stylobate upwards in the middle *(left)*. This corrects the visual distortion that would otherwise occur *(right)*.

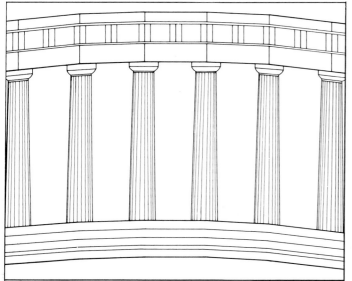
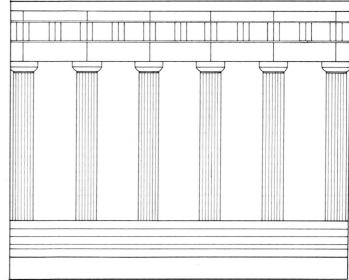

movement in space

Encourage the participants to experiment with movement in space. Ask them to stand at one end of the long corridor: "What action does this corridor suggest? If you were a child, what would you want to do here?" (walk fast, run, skip) "What senses are you using when you run?" Ask the group to find and view a wide staircase: "What movements does the staircase suggest to you? Would you want to run up and down the stairs? to sit on the steps? or to take them two at a time?

"Can you sense how you keep your balance while moving up or down the stairs? What muscles sustain the major strain going up? going down?" (Instructor teeters on stair.) "What did you feel in your own muscles when you saw me off balance? This would be a visual interpretation of a proprioceptive perception or a kinesthetic perception, or both."

preparation

black felt-tipped pens

straight edges

an object with unexpected properties, for example, a plastic salt shaker that looks like metal or wood and is filled with sugar

blindfolds (one for every two participants)

Locate a long corridor to use in "description of partner."

Locate an inclined plane or stairs that invites movement such as jumping.

further exploration

"Individually or with a group play an association game:
If this sound were a smell, it would be _____.
If this taste were visual, it would look _____.
What smells do you remember from childhood?

"Examine the world through magnifying glasses, tubes, colored lenses.

"Create an array of textures for a blind person to enjoy.

"Read the article by Hans Elias and then think of other examples in science that illustrate the need to distinguish between perception and sensory reception."

check list

Did the participant articulate his responses to sensory stimuli?

Was the participant able to distinguish his own associational responses (perception) from his sensory reception?

Did each participant recognize his use of vision to interpret for other senses?

resources

Birdwhistle, R. L.: *Introduction to Kinesics*, Foreign Service Institute, Washington, D.C. (1952). Systematic studies of how individuals perceive, and orient to, one another.

Elias, Hans: "Three-Dimensional Structure Identified from Single Sections," *Science* (1974) Vol. 174, 993-1000. Misinterpretation of flat images such as those visually observed in electron micrographs can lead to errors in judgment of their form.

Environmental Psychology: Man and His Physical Setting, H. Proshansky, W. Ittelson, and L. Rivlin (editors), Holt, Rinehart and Winston, New York (1970). The article by Craik (Pages 646-658) describes methods of measuring comprehension of the environment.

Gombrich, E. H.: *Art and Illusion, a Study in the Psychology of Pictorial Representation*, Pantheon Books, New York (1960). Chapter 7 analyzes how illusions are used in paintings to fool our senses.

Luckiesh, M.: *Visual Illusions*, Dover, New York (1965). An illustrated discussion of the causes, characteristics, and applications of illusions.

McKim, Robert: *Experiences in Visual Thinking*, Brooks/Cole, Monterey, Calif. (1972). Exercises designed to stimulate perception of the visual world.

St. Exupery, Antoine: *Wind, Sand, and Stars*, Harcourt Brace, New York (1949). Sensory descriptions.

Weintraub, D. and Walker, E.: *Perception*, Brooks/Cole, Belmont, Calif. (1966). A concise introduction to visual perception.

Process

Valuing the process of creating: exploration of the characteristics of unfamiliar media leads to recognition that the agent, the medium, and the process are each part of the product.

the medium is one message

The medium influences the product. All media have particular limitations and potentials. "Medium" is the material from which a "product" is made and is itself a means for communicating a message. "Messages" can be conveyed in many media. Words, images, body movements, and materials are all "media" from which products are made. Each of these media can carry a message; for example, a plastic plant can convey a message different from that of the natural plant.

"Process" influences the product and also communicates a message. "Process" is a series of actions carried out with tools that lead to a product. The process of doing can be a very desirable end in itself, for example, the sensory stimulation that comes with kneading dough, sculpting clay, or filming a movie, is pleasurable.

An "agent" (maker or designer) translates a medium by process into an object. Medium and process affect the message the product conveys. A well-designed product requires the agent's commitment to process and requires that he understand the potentials and limitations of the medium. All "products" contain evidence of process, of the tools used, and of the agent's intent and competence, and also convey a message.

Agent, process, and medium inseparably contribute to the over-all impact of every product, including the never finished product, city. The choice of building *materials* and the way they are combined and juxtaposed influences the message of the city — "charming," "sophisticated," "impersonal," "disorienting" — just as they influence the message of the vase.

Building sandcastles is an enjoyment of *process*.

dough sculpture

The impact of this experience is enhanced if the participants do not know what medium is to be used.

"We have worked so far with one medium, cardboard, so let's describe and characterize it." Extend the discussion and ask the participants to comment on the significance of the interaction of agent, medium, and process. "What were the characteristics of cardboard that most helped you in constructing your personal space? What were the most difficult properties of the cardboard to work with? Did you feel you were controlling the material or the material was controlling you? What tools did you use?"

"Now we are going to work with a material that has very different properties. This is the material in its final form." Break up some homemade bread and pass out pieces for the group to butter and eat.

"This is dough ready for you, the agent, to use as a medium." Pass out prepared dough and ask the participants to "Experience this with your senses; only after you have done that, think about it. What are all the different processes that are used to get dough to this stage? What differences do you see in these various flours (examples on hand)?

"As you work with this dough, explore its characteristics as a material. In what ways does it differ from cardboard? Does it have any properties in common with cardboard?"

Ask the participants to each sculpt as tall as possible an abstract form with the dough. As they work, encourage them to try various ways to achieve height with this material. When they are nearly finished, suggest that they work specifically with the surface of their sculpture. They can cut the surface with a knife, glaze it with egg yolk, and add texture with poppy and sesame seeds.

While the bread sculptures are allowed to rise (about thirty to sixty minutes), begin working with the clay.

our experience

Most participants need considerable help in comprehending and verbalizing the inseparability of agent, medium, process, and product in all manifestations of the environment.

It is helpful to put each experience into larger contexts, asking the participants to express in writing as well as orally how the concepts embedded in the activities apply to their larger, everyday world.

The bread dough exercise is so unexpected a class activity that we set the tone of a serious learning experience before we handed out bread to eat.

With some groups, these learning experiences can be intensified by combining them with those of Texture. In this case, introduce the discussion while the participants are working or as they become too occupied with producing "objects." Select texture as the one component of the discussion to explore further. Ask the group to experiment with textures on the bread and clay.

Could you sculpt this piece from bread? from clay? "49" was actually carved in marble by Indian artist Douglas Hyde.

clay sculpture

Ask the group to help themselves to *big* chunks of clay. After they have played with the clay awhile, ask them to shape as tall as possible an abstract form that has sensuous connotations. As they work, help them verbalize the process. When someone has completed a form and you have discussed it with him, ask him to make another from the same clay.

As they work, ask the participants to compare clay, dough, and cardboard as media. "Compare the processes used in working with each. In what ways did you use your hands as tools with each medium?"

Pass around examples of finished bread and clay sculptures and the raw ingredients of each. Encourage participants to experience and play with these materials to help them better analyze the sculpted objects. Now return to the bread sculpture. When the dough has risen, let the group watch it bake. Provide butter so everyone can eat bread when it is finished. This emphasizes that often the process, in this case that of making and eating, is more important than the "product."

spider web

Ask each participant to bring to this session a large roll of string. Using the entire volume of the room, construct a spiderweb. In the discussion, point out examples of the many things that help define space: color, texture, pattern, order, and process.

"How do string and clay and bread differ in their sensory stimulation?" (odor, texture . . .)

discussion of clay and dough

List attributes suggested by the group as being common to the objects. Analyze the attributes of several objects in each medium by mass, height, shape, volume, odor, space, color, and texture.

"Find adjectives to describe each object in terms of its characteristics. Is it round or square, thick or thin? Distinguish between direct sensory experience and visual interpretation." Ask the group to compare clay and dough with cardboard, a medium they are already familiar with.

"Which is the most important: medium, process, or agent?" (It's impossible to say.) Discuss the concept that the product is inseparable from the producer; the process, from the medium.

"In what ways does one's choice of medium affect the message? What are the limitations of the selected media? What advantage did one medium have over another? Why did we include bread dough? How did the process affect the product? What would have been the differences in your personal space if it had been built of clay? cloth? wood? Is there any value in process for its own sake?"

Extend the discussion to include the participants' everyday environments. First, analyze the room in terms of the media, the agents, and the processes used to produce its components. Then consider the building and site as a whole.

"Can you apply these simple experiences to even larger problems? What are we learning about ourselves in making our civic environments? Name some." (We build freeways because we value fast movement.)

"Feather Woman" prepares a base for firing her pots by layering sheep-dung coals and broken pots.

"Feather Woman" lays her bowl, which has been shaped, polished, slipped, and decorated, in place.

She covers the bowl with pottery shards to protect it from ashes and flames.

"The clay is a living being when you put it in your hand, you know. Look at it. A lump . . . a lump that says to me, 'Make me as I am . . . make me beautiful.' So we converse every step of the way, the clay and I."

Elizabeth White from an interview with Maggie Wilson, *Arizona Highways*, May, 1974

"Modern design should express the qualities and beauties of the materials used, never making the materials seem to be what they are not.

Modern design should express the methods used to make an object, not disguising mass production as handicraft or simulating a technique not used.

Modern design should master the machine for the service of man."

from "Twelve Precepts of Modern Design," *What is Modern Design?*

**clay and plastic;
two media, two processes**

A high-speed thermoformer continuously molds polypropylene and recycles its own scrap.

The air-cooled molded sheet emerges as trimmed margarine tubs or juice containers.

preparation

previously prepared examples of objects sculpted from clay and bread dough

Bread sculpture: homemade bread, prepared dough or 1 loaf per participant of frozen bread dough (for example, Bridgford Frozen-Rite), flour with which to work the dough, samples of different kinds of flour and cornmeal, egg yolks and pastry brushes for glazing sculpture, poppy and sesame seeds for texture, pie pans or cookie sheets for sculpting and baking, ovens (sculpture can be satisfactorily carried from one building to another), knives, butter, napkins, sponge, and paper towels.

Clay sculpture: one or more kinds of mixed clay, sample quantities of the clay ingredients, cardboard for each participant to work on, newspaper-covered worktables, a few clay working tools, rags or paper towels, and a bucket of water. (The tables must be thoroughly cleaned so that no toxic materials from the clay get mixed with the dough.)

Save examples of textured clay pieces to use in Chapter 7, Texture and in Chapter 8, Light.

balls of string

further exploration

Ask each participant to write a page on what, specifically, he thinks were the concepts behind this session. "Give two detailed examples of how these ideas apply in your city."

Suggest selecting a medium with which the group has not worked and creating a sculpture.

Explore the "process" of real experiences: for example, how a car wash operates or how a McDonald's hamburger is made.

There are various ways to give the abstract concepts of this chapter concrete reality in the context of a specific subject. For example:

Science
   "How does the scientific method handle 'agent' and 'process' in order to observe, define, and measure properties of materials?"
Mathematics
   "How could the sculpted forms be measured to find their surface areas, volumes, and centers of gravity?"
Social Studies
   "Give examples of ways in which art forms have been used in political movements in the twentieth century."
English, Reading, Film
   "In what ways does a writer or photographer communicate his own direct sensory experience to others?"

check list

What evidence did the participant show that he recognized the value of process for its own sake?
Does the participant more clearly understand the inseparability of agent, medium, process, and product in all aspects of environment?
Was the participant able to gain further confidence in understanding and imposing his own will to create order?

resources

Heimsath, Clovis: *Pioneer Texas Buildings — A Geometry Lesson*, University of Texas, Austin (1968). Photographs and sketches relate buildings to simple geometrical solids and voids.

Johnson, I., and Hazelton, N.: *Cookies and Breads — The Baker's Art*, Reinhold, New York (1967). The Museum of Contemporary Crafts illustrates the use of dough as an art medium.

McLuhan, Marshall: *Understanding Media*, McGraw Hill, New York (1964). A provocative book by the author of the phrase "the medium is the message."

Masters of Modern Art, A. H. Barr, Jr. (editor), The Museum of Modern Art, New York (1958). Distributed by Doubleday. Sections on photography, film, design for the theater, architecture, and crafts.

Oka, Hideyuki: *How To Wrap Five Eggs*, Harper and Row, New York (1967). Japanese design expressed in traditional forms of packaging.

Read, Sir Herbert: *The Art of Sculpture*, Pantheon Books, New York (1964). An esthetic view of the art of sculpture.

Structures, Elementary Science Study of Education Development Center, Inc. 55 Chapel St., Newton, Mass. 02160 (1968). Pamphlet giving inspiration, materials, techniques, and examples of constructions out of clay, straws, and paper tubes.

5

Order

Organizing sensory data into meaningful patterns: participants begin to develop
the ability to evaluate order in both the built and the natural environment and
to comprehend the function of all order.

the tendency to see patterns
is in the brain—not the eye

We organize the sensory data we receive into meaningful wholes to help orient
ourselves in the environment. If there is no ordering, no synthesis into a whole,
the data have no meaning. We are left with a feeling of dissonance that is a
mild form of chaos. We have no reference point for action. We seek order to
reduce our anxiety about seeming chaos, so we can know where we stand.
When simplified, the data become more memorable and we can more easily
manipulate them mentally. In the forest we need clearings, special trees, or
paths to help us find and remember our way. In office buildings we rely on
sequentially numbered doors to orient ourselves along endless corridors.

On the other hand, it is more interesting and involving not to simplify our
world too far. Tension can be deliberately created by retaining some degree of
complexity. A composer organizes a trite theme of a few notes into patterns
that, by their variety of rhythm, ornamentation, and orchestration, surprise and
delight our ears. The degree of complexity we prefer constantly changes. Styles
in everything we design swing from Victorian "clutter" to contemporary
"starkness" and back.

If past ordering has been too rigid and tight, future material cannot be
projected into it and therefore sensory perception is blocked. Over reliance on
"stereotypes" is an example of too rigid an organization of data. Anything
"new" is rejected. A new skirt length is greeted by open-minded fashion
pacesetters with delight; by others, with derision. In time, however, the new
length becomes a familiar pattern, accepted by all.

We relay on perceptual habits to establish order out of chaos. In spite of the
bombardment of information we continuously receive through our senses, our
world does not seem to us totally chaotic. This is because we unconsciously use
mental tools to establish order in our environment.

We use line to give definition to infinite space. Line can divide, encircle, box
in, or shut out space, establishing *object* and *field*. The lines defining the
Andromeda constellation are our invented device for organizing that particular
portion of our visible universe. Andromeda becomes an *object* distinguishable
from its *field*, the sky.

**we organize data into mean-
ingful wholes**

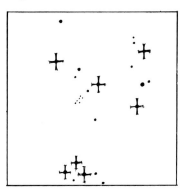

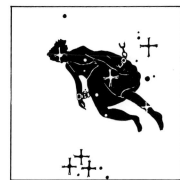

we group by similarities

We use such attributes as size, shape, value, proximity, and direction to help us sort the jumble of incoming information into manageable categories of "big" things, "square" things, "dark" and "light" things.

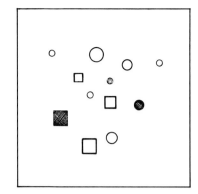
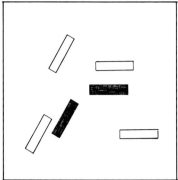

we extend patterns beyond what our eyes record

Rhythm lets us understand many parts by inference from one part, which we understand already, without looking at every part in detail. Relying on the symmetry and rhythm of patterns previously learned, we "fill in" the picture. The mind embellishes what the eye sees.

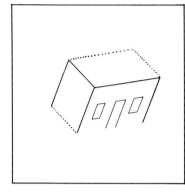
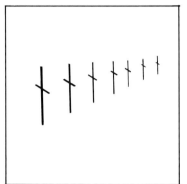

we seek out familiar patterns

Sometimes we are fooled. In recognizing patterns or grouping by familiarity, the brain "reads" more than is seen by the eye. We seek out pattern even though the pattern may be unintended. We register the pattern "house," because it is more familiar to us than the pattern "HQLIS5."

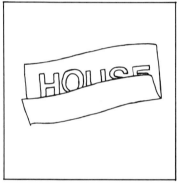

we both simplify and elaborate what is "real"

We need to orient ourselves in our environment, but we also need sensory stimulation. Thus, when we make our own representations of reality, we both eliminate part of what is real, and add details that are not. In the cartoon, Nixon has been captured by a single feature, his nose; John Q. Public's confusion has been expressed by drawing four heads.

ordering random lines to make them memorable

Hand out two sheets of paper and a broad-line pen to each person. Ask each participant to "Draw one line on your paper. There is only one condition that must be met: the ends of your line must begin at one edge of your paper and terminate at one of the other three edges."

Divide into groups of five or six. Ask each group to "make order," that is, an array, out of their papers. (Possible solutions will include geometrical patterns, a straight line, or even the arrangement of all the papers into a stack.)

Ask each group to visit all of the other group arrays. Then ask each person to draw from memory (without referring to them a second time) as many of the arrays as he is able. With the whole group, compare these sketches with the original arrays still in place.

"Which arrays proved easiest to remember? What makes them memorable?" (In general, the most highly ordered pattern will be the most memorable.) Still referring to the original arrays:

"Which array do you like the best? Is it also the one that was easiest to remember? Why or why not?" (Perhaps less order is more interesting.) "When you look at the whole, which do you see, an 'object,' or the 'field'? Which has the most content? Does more information make a more interesting array? Can there be too much information?"

Mix up the arrays of papers and collect them into one pile. Throw down one paper at a time and afterwards ask the group if they can remember the number of papers and the configuration of each line. (There is an order but there are so many variables we cannot perceive any order. Thus we are disoriented and have difficulty remembering specific patterns of lines.)

strengthening the focus of the course for the participants

We emphasize ordering as it relates to the disciplines of the participant group. For example, *line* can be defined from different points of view — math, art, dance, or woodworking.

Participants who are accustomed to course outlines often have some difficulty recognizing that they are learning in a course without lecture notes. You can provide structure and closure with questions. For instance, "Reflecting on the use you have been making of all the senses, what do you think this course is about? How does one order things? How do we keep our sensory receptions from bombarding us until we can no longer function? Can you 'order' the experiences we have had today? Can you extrapolate from these trivial experiences to solve larger problems? Of what value would these experiences be in training a scientist? a carpenter? a psychologist? a business-person? a housewife?

"How do you think you learn? What is the process? Isn't learning largely the ordering of information into relationships? This increases your insight or understanding and also makes it memorable and therefore retrievable. Of the things you are learning now, how much do you expect to remember twenty years from now?"

ordering for orientation
two-dimensional representation

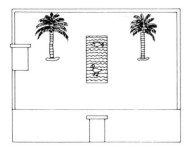

"Mappers of the city must also be aware of how to use design tools (such as line and pattern) to make a town or city comprehensible. The map maker selects, that is abstracts, those pieces of information that he thinks will best send his intended message. A map is two-dimensional; the viewer therefore has a bird's eye view of the city. "Imagine a bird's eye view of a town or city with which you are familiar. Select the information necessary for an automobile driver to get a general picture of the layout of this city, and draw it on a piece of paper.

"You have, with line and pattern, abstracted and ordered information so that someone else can orient himself and move around. Look at the different street patterns drawn by one another. Can you divide them into two or three basic categories of patterns?" (The instructor may add any missing category, such as radial streets.) "What tools are used in these two-dimensional representations to help you know how to get from one place to another? Which street city plan is easiest to remember? Why? Which looks the most interesting? Why?

The drawings above illustrate the problem of portraying three dimensions on a flat surface. The *upper figure* is our modern way of representing the garden space portrayed in the ancient Egyptian way *below*. There is more information in the ancient way. We would have to draw both a plan and a perspective to achieve the same communication. Note the way Egyptians portrayed vertical height for walls and two doorways.

Melanesians used sticks, representing currents, and shells, representing atolls, to orient themselves as they charted the South Pacific.

three-dimensional reality

"In reality, we move at various rates of speed walking or riding through a three-dimensional city. We need to make order out of this jumble of buildings, streets and freeways so that we can move around knowledgeably.

"Imagine you are lost in your city. What clues are there that you could see, hear, or smell (tall buildings, church bells, bread factory) that tell you, you are downtown, or at some other particular location?

"List all the things suggested that help people place where they are and know how to get where they are going. Include some famous landmarks. Now, using your mind's eye view of yourself experiencing your city in three-dimensions, annotate on your two-dimensional map, the additional clues you need for comfortable, constant orientation in your town."

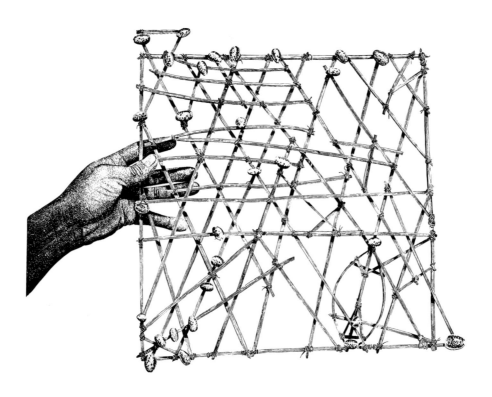

ordering for sensory stimulation

Ask everyone to draw one member of the group. Give a five minute time limit. Discuss together what attributes of the person have been abstracted: arms, legs, hair.

"What details are common to all your drawings and which details have been represented by only one or two of you?" Show the participants examples of figure drawings which vary noticeably in their amount of detail. (Choose, for example, drawings by Wyeth, Picasso, and Matisse.)

"We have looked at abstractions you each made. Now look at the abstractions these artists have made. The artist abstracts, just as the map maker does, from a whole. But what he chooses to select and abstract and how he chooses to represent that abstraction, makes all the difference. The sculptor orders by eliminating extraneous detail. The painter abstracts a three-dimensional object into a two-dimensional representation. The artist works in a range of abstraction between photographic duplication and the child's representation of a person as a stick figure.

"The artist picks lines that convey the essence of the figure. He brings out not only the identification of the subject, but also its particular qualities. How well the artist succeeds in conveying the essence of his subject is a measure of his greatness.

"Whether you like or understand the work depends on your own previous experience and understanding. If, in your mind's eye, in the past your sense of ordering has been very tight, that is, you always like the same thing, then when you come across new ways of abstracting and ordering, they have little meaning for you. Very often when faced with an unfamiliar ordering, one rejects it outright. It is more difficult and certainly more time consuming to try to understand a new way of abstracting, before deciding whether or not it is esthetically, emotionally, and intellectually pleasurable to you. But it often is very satisfying to engage in this critical process.

"We need to find order to simplify data so that it can more easily be mentally manipulated. Simplified, it is more memorable. On the other hand, by introducing more complexity, tension can be deliberately created to heighten interest. Compare, for example, drawings by Matisse with those by Picasso.

"Consider how patterns of line, symmetry, form, balance, and so forth, are also used to establish order in dance, music, literature, sports, poetry, and the theater. In considering a particular work, imagine it revised with greater and lesser degrees of order and more and lesser amounts of information content. Be able to express an opinion and rationale for your opinion.

"How do we order other kinds of information? In mathematics, science, history . . .?" (We express complex relationships in simple equations. We organize historical events into "cycles" and "trends.") "What examples of order do you find in nature?" (patterns of leaves and branches)

photographer

child

Kuniyoshi, Self Portrait as a Golf Player, 1927

Picasso, Card Player, 1913-14

Lipchitz, Figure, 1937

preparation

unlined 8½ inch by 11 inch paper

black *broad-line* felt-tipped pens

a variety of line drawings by artists such as Matisse, Klee, Picasso, Kandinsky, Wyeth, and Japanese sumi painters

further exploration

Show the group a drawing of a Picasso dove. "What are your associations with the picture?" (communism, peace . . .) "What do these associations have to do with your reaction to a work of art? Do they hamper you in forming a value judgment of it as art?"

"Analyze the kinds and number of decisions that had to be made in creating the two-dimensional order that is this book. Analyze materials as well as form (pattern)."

check list

Was the participant able to use such tools as line, pattern, form, symmetry, and rhythm to establish order?

Was the participant able to articulate why we need both order and continual sensory stimulation?

resources

Arnheim, Rudolf: *Art and Visual Perception, a Psychology of the Creative Eye*, University of California Press, Berkeley, Calif. (1974). The first four chapters cover in detail the concepts introduced in Order.

Giedion-Welcker, Carola: *Paul Klee*, Viking, New York (1952). A chronological anthology that includes some works of Miro, Picasso, and Braque.

Gombrich, E. H.: *Art and Illusion, a Study in the Psychology of Pictorial Representation*, Pantheon Books, New York (1960). See Chapter 8, "Ambiguities of the Third Dimension."

Heimsath, Clovis: *Pioneer Texas Buildings — A Geometry Lesson*, University of Texas, Austin (1968). "Ordering" as exemplified in simple, small buildings.

Kandinsky, Wassily: *Watercolors, Drawings, Writings* (translated by Norbert Guterman) Abrams, New York (1961). Black-and-white and color reproductions of Kandinsky's work.

Lieberman, William: *Matisse: Fifty Years of His Graphic Art*, Braziller, New York (1956). Exposition of Matisse's use of line in his art.

Masters of Modern Art, A. H. Barr, Jr. (editor), The Museum of Modern Art, New York (1958). Distributed by Doubleday. Illustrations of a variety of artists' styles.

Paul Klee: The Thinking Eye, Jurg Spiller (editor), Wittenborn, New York (1964). The definitive source on Klee includes good examples of line drawings.

Sabartés, Jaime: *Picasso: Toreros*, Braziller, New York (1961). Black-and-white line drawings and lithographs.

Tanaka, Ichimatsu: *Japanese Ink Paintings — Shubun to Sesshu* (translated by Bruce Darling) Weatherhill (1972). A survey of Japanese sumi painting.

Two Worlds of Andrew Wyeth: Kuerners and Olsons, Metropolitan Museum of Art, New York (1976). Many pencil and dry-brush studies leading to Wyeth's paintings.

Wheeler, Monroe: *The Last Works of Henri Matisse*, The Museum of Modern Art, New York (1961). Color and black-and-white plates of Matisse's large cut gouaches.

Structural order

6

Finding order in structural patterns: a study of the environment by drawing on past experiences and building a simple structure leads to an understanding of the function and the visual affect of structures.

do sky hooks hold it up?

Buildings, furniture, towers, or walls and plants are all held up in similar ways. We have the "post and beam" desk or house roof; the pedestal chair or umbrella roof tent; the cantilevered street light, the cantilevered tree limb, or the overhanging balcony; the spider web and the suspension bridge. All express particular patterns of structure.

When we make and build we are structuring, we are ordering. To shape a building is to create order in three dimensions. The structures we use may be adapted from nature or from previously built forms, or they may be newly invented.

We use the word "structure" to connote many things. In built environments the word means the orderly relationship of parts. Structure also connotes the functional means of spanning space. Our sense of well-being depends upon the care with which the spanning of space is executed. Structure is the means of making a material stand up to defy gravity. The steel frame of an office tower must support itself in space against wind, and must support a live load of people, rain water, and other things set upon it. "Structures" in this sense must *appear* firm and safe as well as be so. No one wants to be housed in a structure that *feels* as if it may fall down. Hence we may think of a "visual" structure that may or may not be synonymous with the functional one. In early times it was synonymous. Stone walls and stone-vaulting were the finished surfaces as well as the means of support. Structures still tend to be revealed or expressed in buildings, but they may also be concealed because of various constraints or by preference.

The Greeks could span some 40 feet with their stone post and beam structures. The Romans increased the distance to 150 feet using brick masonry arches. We are able to span more than 5800 feet with tensile structures.

A discontinuous compression structure proposed by Buckminster Fuller is here developed as an art form by sculptor Kenneth Snelson. The parts each express the work they do. Cables in tension hold each end of each pipe fixed in space relative to the others; the pipes are in compression. Unlike traditional gravity structures that will only stand up when perpendicular, this particular tensile structure will stand by itself at any angle to the earth. The view *above* is from the outside. The view to the *right* shows the inside of the structure, looking up.

The technology used and the available materials with their particular properties, are determinants of what is built. With today's technology we have developed buildings whose "skins" are entirely separate from their structures. New technology frees us from our accustomed patterns of putting things together. We create structures that span the widest bays and that rise higher than a redwood tree.

We live in an era of such rapid transition that our sense of what is secure and what is pleasing has difficulty keeping pace. Some of us long for a house rooted in the earth by a basement or for thick, stable-looking overhead beams. Hovering cantilevers, cables, and glass walls may excite our imagination, but require adjustment to a different sense of security. When the Eiffel Tower rose in Paris, many declared it hideous and unsafe because it was so different from the stone buildings they were accustomed to. Our perception of many modern buildings is clouded by our incapacity to feel their stability. The structural glass railings leading across a moat into an Expo 67 pavilion were avoided, even though the glass was an inch thick. No one would walk near them for fear of falling over the side. It takes time to become comfortable with new structures.

on the left the palm leaf is stiffened by what we call a folded plate principle and cantilevered from its stem.
on the right a roof is held up by steel beams cantilevered from a central column.

drawing on memory

"Think about the structures you have built so far. What physical forces can you identify that help explain why you were able to build a structure of clay as high as you did? Why did it tend to fall down? How much did you change the original design of your personal space because you couldn't get it to stand up using the materials and skills you then had? How did the feeling you wanted to convey change because you had to change the structure to make it stand up? In how many personal spaces you visited did their apparent lack of stability detract from their affect on you? How many spaces achieved apparent stability because of their simplified post and beam structure? How many seemed stable only because they hugged the ground?

"There are external forces which affect any structure's ability to stand up, as well as the internal forces of the structure. How high could you theoretically build a tower? a sphere? Why? How high could you build a rectangular building on the outskirts of Chicago? What external forces might limit you?

"The process used, the materials, the degree of skill, and the design ability of the builder all affect the outcome of the product. In your experience with cardboard, which of the preceding variables had the greatest influence on the outcome of your personal space? Which was the second most influential variable?"

searching for structures

"The physical principle of opposing forces and its two corollaries, tension and compression, can be used to describe why our structures stand up. Think up and do as many simple demonstrations of these concepts as you can. (For example, draw a corrugated potato chip through some dip with its corrugations perpendicular to the surface of the dip. Does the chip break? Would a plain chip hold better? Why? Slowly open an umbrella. Roll a sheet of notebook paper into a column that will support a tennis ball. Suspend a book from a string.) Explain the physical principles as you demonstrate each example.

"Use the immediate environment and identify structures — chairs, desks, plants, walls, doorways, roofs, buildings, and trees. Point out and describe how opposing tension and compression function in each case. What external forces must be taken into account?" (The weight of people, rain, snow, a roof; a high wind; the structure and relief of the ground underneath.) "Now analyze the form, volume, and mass created in each instance. How does the whole structure affect you as you experience it? What is particularly pleasing and why?

The bridge over this canyon is made out of paper.

"Is it old and familiar? Is it new and exciting because of the challenge it presents? Is it interesting; does it ignite the imagination? What structures depend on modern technology, engineering, and materials? What structures, or parts of them, could have been made by builders of the pre-industrial era? Which kinds of structures span the greatest area most efficiently, that is, with the lightest weight?"

holding things up

Divide the participants into teams and ask each team to design and build a structure that will support at least one hundred pounds. Use wood, cardboard, string, rope, or any other readily available material. (Paint cans, cardboard boxes, or other ready-made objects are not acceptable.) When they are finished, ask the participants to analyze the structures and explain why they can hold up so much weight. Then analyze each structure's form, volume, and mass as if the structure were a piece of sculpture to be enjoyed only visually.
"Now analyze the structure in terms of how it functions as a seat. Could this seat have been constructed in the 1800's? What is there about this seat to remind you of one you have seen before? What is reminiscent of structures found in nature? In what type of place would this seat look best? function most appropriately? Why?"

cardboard

matte knives

duct tape

string

materials to demonstrate structural principles such as an umbrella, wrinkled potato chips and a dip, tennis ball

This is a good occasion to invite an architect and a structural engineer to meet with the group.

further exploration

Ask participants to visit a favorite building and explore why it stands up. "Is the structure of the building visible? expressed, but not actually visible? concealed? What materials are used? How does the structure contribute to the visual success of the building? the functional success of the building? Does the building *feel* strong? Photograph and diagram the structure of the building.

"Find three structures in the botanic world that suggest man-made structures you have seen. Do you perceive how the botanic structures were used in fabricating the man-made structures? Diagram first the botanic structures, then the man-made ones. Compare the diagrams."

check list

Were the participants able to build functional and pleasing structures?
Were the participants able to express why those structures worked?

resources

Huxtable, Ada Louise: *Pier Luigi Nervi*, Braziller, New York (1960). A highly readable monograph on this inventive Italian engineer.

Macaulay, David: *City*, Houghton Mifflin, Boston (1975).
Cathedral, Houghton Mifflin, Boston (1973).
Castle, Houghton Mifflin, Boston (1977).
Pyramid, Houghton Mifflin, Boston (1975).
Underground, Houghton Mifflin, Boston (1976).
Line drawings reveal the building process.

Salvadori, Mario and Heller, Robert: *Structure in Architecture*, Prentice-Hall, Englewood Cliffs, New Jersey (1963). A book for both layman and architect written in literary, rather than technical language.

Wandering Wall, a film available from the University of Iowa Audio/Visual Center, Iowa City, Iowa 52240. Describes the Best showroom.

Other chapters also contain examples of structural order:
Post and lintel (beam) structures: Stone — page 32, page 147 (the beams of the Turin arcade span from a wall of piers and lintels on one side to a wall of masonry arches on the other). Wood — pages 43; 70; 91; 144; and 154 (the 1873 City Hall). Steel — pages 69 (pipe braced by diagonal structural members in tension); 83; and 124 (wide flange cross-sections).

Trusses: Steel — pages 83 (the lightweight bar joists supporting the ceiling are the most common steel commercial truss); 131 (Which looks safer, the side view showing the trusses, or the cross-section revealing inadequate steel arches?); and 146 (concealed trusses span 150 feet to hold up the roof).

Arches: Carved in natural rock — page 70. Built into stone-bearing walls — pages 25; 68 (a new post-and-lintel skeleton wall with in-filled masonry sits on a traditional masonry bearing wall with arched openings); 70; and 97. A relieving arch following the lines of stress placed on the wall which supports a dome out of the photograph to the upper left page 58. A concrete structure sheathed with a masonry arch for appearance only — page 130. An arch 600 feet high, hollow, and triangular in cross-section — page 75.

Domes: Built with steel trusses concealed from both the interior (page 65) and the exterior (page 73). A masonry dome held up by half-dome vaults visible in the wall below — page 76. A decorative steel sphere supported by concrete columns made by slip-forming, a process that influences both the shape and the horizontal texture of the columns — page 133.

Cycloid shell: Of reinforced concrete and post-tensioned steel (it is 96 feet long) — page 59.

Cantilevers: Of steel-reinforced concrete, making possible the stepped-out, sloping sides — page 155 (the 1978 Dallas City Hall); and *left*.

How do you react to this structure? The entrance slides open by means of hydraulic jacks. A hidden structure of cantilevered beams holds the masonry in place.

Texture

7

Interpreting tactile stimuli: comparison of small-scale textures of objects with large-scale textures of the urban environment serves to introduce texture as a tool for orientation.

we "touch" with our eyes

Texture is an important component in the total environment. It is perceived directly by touch and secondarily by visual interpretation of past tactile experience. We have experienced the soft texture of a sweater or the rough texture of a brick both directly by touch and indirectly by visual interpretation. When we speak of the texture of a neighborhood or city, though we can feel, hear, and smell it, we are usually referring to a secondary experience of visual interpretation.

Texture always involves three-dimensional ordering on a surface, whether it is on a small scale, like the wood grain in the dining room table, or on a grand scale, like waves on the sea, sand on the beach, and summer homes on the shore.

Texture is used as a tool for orientation. We expect to see textures at an accustomed scale. When we do not, we are momentarily disoriented. This applies equally to the small-scale texture of an object and to the large-scale texture of an urban environment. For instance, the white two-story clapboard surfaces of a New England salt box house would be a startling texture among the low adobe-walled houses of New Mexico.

Texture is ordering that can also be used to provide enrichment for variety and relief from tedium. Conversely, of course, texture can be used purposefully or accidentally in so much variety that it evokes feelings of chaos.

Selective inattention is used to insulate oneself from abrasive texture. One tries to ignore a scratchy sofa or the clamor and chaos of a busy street. Selective inattention also insulates us from the monotony of textures that are too uniform.

The baby touches with her hands. As she grows, she learns to "touch" with her eyes, like the adult.

qualities of textures

Encourage the group to view and handle a variety of textures common to both natural and built environments. Assemble in small groups and articulate the qualities of each texture, one at a time.

Ask each person to experience the texture visually from a distance and describe it. Ask him to use as many descriptive adjectives as possible. Record the adjectives. Then ask the participants to experience the texture tactilely, perhaps first with their eyes closed, and again record their descriptive adjectives. "Which list is longer? How do you account for this?"

If the group members are having difficulty bringing additional adjectives to mind, ask them to complete the sentence: "If this texture were a person, he or she would be _____." Repeat the exercise, listing the additional adjectives.

"How could you use this texture? Where would you expect to find it in the environment? Is it pleasant to touch?"

"Imagine yourself in a room, entirely surrounded with a particular texture, for example, straw or velvet. What is your emotional response to this environment you are imagining? Do you associate it with anything in your own past experience?"

creating textures

Demonstrate the process of creating a textured surface using clay. Have the participants select three effects (such as touchable, hostile, soothing) and make three corresponding textures with clay. Different tools can be used, such as hands, a comb, and sandpaper.

"Try a directional light on your texture, first straight on, then entirely from the side. How does the light alter it?" (Sidelight will make a heavily textured surface seem much bolder with the contrast of light and shadow.) Referring back to the concepts in the chapter Process concerning agent, medium, process, and product, discuss: "Does the process by which this texture was created influence the product? Why and how? Does your own involvement in the process of making the texture add personal value to the object for you?"

texture in the environment

Take a walk outside, around buildings or a campus. If possible, walk down a street with a commercial strip and through a planned unit of several buildings, such as a shopping mall or an apartment complex. The purpose is to identify and study the intended use or accidental happening of both man-made and natural textures and their combinations.

"Identify the textures used on building surfaces. Are the textures used in such a way as to be boring if seen on a second trip? Are there too many textures on any of the buildings? What are the natural textures you see? Are there combinations of natural and built textures? How well do they work together? Give examples of what might work better.

"Do you see examples of how the architect has used texture to create shadows? Find examples of textures that distract from or contradict the intended use of the space. An example of an inappropriately used texture might be splintery wood benches in a shopping mall.

"What texture do the buildings themselves create in the city? Describe two cities that have a different 'feel' in terms of texture."

These two walls, one modern, the other ancient, each use texture for a specific effect, which is heightened by the play of light on the surface.

preparation

a variety of natural and man-made textured materials or objects

cardboard for each participant to work on

prepared clay

clay working tools

materials for cleaning up

a high intensity flashlight

Save examples of textured clay pieces to use in Light.

further exploration

"Discuss the use of 'texture' in music. Compare, for example, the same theme played by strings, by horns, and by a harpsichord. Or compare classical, electronic, and hard rock music.

"How does the texture of music interact with the texture of built environments? What textures are needed in building surfaces to augment different sound qualities? What textures are needed to augment the spoken voice?"

check list

Did the participant fluently articulate his feeling responses to texture? Is he becoming aware of how positive or negative associations from his past influence his response?

Was the participant able to analyze and describe a texture?

Did the participant achieve his desired effect when he created the texture?

resources

American Craft: Each issue of this monthly periodical illustrates the use of texture in metal, fiber, ceramic, and other crafts.

Broadatz, Phil: *Texture,* Dover, New York (1966). Consists of 112 plates of photographed texture.

Branch, Daniel Paulk: *Folk Architecture of the East Mediterranean,* Columbia University, New York (1966). Preindustrial environments often contain readily discernible textures because of their handcraft technology — so different from our own.

Collins, George R.: *Antonio Gaudi,* Braziller, New York (1960).

Conrads, U., and Sperlich, H. G.: *Phantastische Architektur,* Verlag Gerd Gatje, Stuttgart, Germany (1960).

Gibson, J. J.: *Perception of the Visual World,* Riverside Press, Cambridge, Mass. (1950). An analysis helpful in furthering our understanding of the way we react to texture.

Light

The role of light and color in perception: light and color can be manipulated to demonstrate their role in ordering visual stimuli and evoking response to the environment.

"colors are the children of light"

The visual world exists only because there is light. "Light" is the medium through which all visual stimuli are carried. What we call "color" is light waves having specific energies. The direction, type, and intensity of the light on an object have as much to do with how we perceive it as does its form, mass, or color. Unlike shape, texture, and other elements of design that can be experienced with other senses, color normally cannot be perceived without light, and so cannot be considered as being separate from light.

Light/color cannot easily be taken out of a particular context for discussion. However, in common usage we talk about "light" and "color" without ambiguity. Thus we can make observations about the effect they have on us in specific instances and the role they play in our perception of the world around us.

We tend to be far more aware of the psychological impact of light/color than of its physical impact. We are only beginning to recognize how essential light is to life itself. For example, it controls metabolic and glandular functioning. The physical reaction as far as vision is concerned is primarily the impact of electromagnetic radiation on the end organs of our eyes.

Our psychological response, however, is triggered by light/color combined with many other variables. A black room is cozy if it is a warm movie theater; it is scary if it is a cold, damp cave.

Directional light articulates the forms of the desert landscape revealing its texture. At high noon when the sun is overhead, the canyon appears flat and colorless.

light can change texture

Using some of the clay textures saved from the session on Texture, ask one group of participants to arrange them on a surface near electrical outlets. "Set up two spotlights or theatrical lights. Beginning with the white directional lights, change first the direction and then the intensity of the light. Observe and discuss how altering the light affects your perception of the textures. Then, trying different colored gels over the lights, vary direction and intensity on the textures."

Cover the interior large surfaces of one of the personal spaces saved from Sequence of Spaces with crumpled white butcher paper. Attach crumpled tissue paper to the interior walls of several additional spaces, using a different color of paper for each space. "Illuminate the first space with a white directional light. Compare the effect of directional lighting on texture at this scale with the smaller scaled texture in clay. What are some of the differences?

"Alter the intensity of the light and note changes perceived in the texture. Step inside the space. Does the visual texture you have added make the space seem smaller or larger? What affect do you think the scale of this texture would have on you if you were standing in the middle of an enormous space? Now enter the spaces with the crumpled tissue paper, each illuminated by a directional light. Which color makes you feel the most comfortable in a space of this size and texture? What color seems the most discordant?"

Light and color support and evoke particular moods. Bright color and soft incident daylight controlled by shutters set a gay lunchtime mood. Architect Warren Platner shifts the color to a warm glow with deep shadows at night encouraging intimate conversation and lingering.

Note the use of scintillation — the deliberate sparkle reflecting off silver, glass, and other metallic surfaces to give a rich, controlled stimulus to the eye.

Compare the dappled pattern of light on the nighttime floor with the pattern of light on page 17.

color can change mood

Choose a semicircular or three-sided space (perhaps one saved from Sequence of Spaces) and make the interior white. One way to do this is to tape white butcher paper onto the walls. Ask the participants to experience the color change within the space by using lights and colored gels to change the interior successively to red, yellow, and blue.

Discuss and record on the board (1) the feelings associated with primary colors and (2) how altering the color changes the feeling of the space or volume. Ask each person to account for the personal biases associated with his responses to these colors.

Ask one of the participants to sit in the space. Light his face with, in turn, red, blue, and yellow. "How does the color affect your feelings about the person's character or mood? Can merely changing the light, with no change in facial expression, make him 'brooding,' 'happy,' 'angry'?"

Reassemble this group of participants for discussion. "What colors besides the three primary ones do you react to strongly? Can you analyze why you have such a strong response? How and why do the reactions of one person differ from those of another? Why are so many reactions similar? Are there both pleasant and unpleasant personal associations with each color? Give some examples."

How do you appear at home, at play, at work, and on the freeway? Compare the hair, skin, eyes, and fabric colors under *(from left to right, top)* daylight, warm white fluorescent light, cool white fluorescent light and *(from left to right, bottom)* incandescent light, yellow filtered light mixed with incandescent (similar to sodium vapor light) and pink filtered light.

color changes perception

our experience

The three learning experiences can be carried on simultaneously in different parts of the room. When a group has finished its experience at one place, it can exchange places with another group. To focus the final discussion, it is best to bring the group back together formally.

The participants enjoy and get a great deal from experimenting with ideas of their own at each of the three stations.

If individuals have difficulty articulating these responses, ask them to make direct color associations such as:
"Black" reminds me of _____.
"Red" reminds me of _____.

Right the bold pattern of black and white marble reveals the underlying structure of the Orvieto Cathedral, but the striping nearly conceals its form. Light both articulates the "real" form of the Kimbell Museum *(opposite page)* and generates additional changing, ambiguous forms.

Ask another group of participants to intensify the emotion evoked in a personal space built in Sequence of Spaces by adding color. Suggest first experimenting with the lights and gels to see if they can intensify even further the emotion being projected by the space or produce an entirely different affect. "What qualities do hard and soft (reflected light) shadows add to your result? What colors are the shadows?"

Ask the participants to experiment with contrast. "Experiment with various color combinations. Notice how the choice and relative amount of a second or third color alter your perception of the first color and of the whole. Note the effects of differing light intensities and of the position of the light source.

"Carry out these experiments with each other acting as 'visitors' and 'interviewers.' When you interview your visitor, try to learn from him what part, if any, of his response he is bringing from his personal past experience. Note what part of his response appears to be a 'universal' reaction."

light changes space quality

After each of the groups of participants has completed all of the projects, ask everyone to reassemble for discussion. "Think of your favorite room. How many different kinds of light are in it? From what directions are the sources? How does the light alter the space? Does the character of the room change if the light changes? Is it a 'daytime' room or a 'nighttime' room or both? How is the dominant color of the room affected by the other colors? For instance, a room that is eighty percent yellow, ten percent orange, and ten percent green will have an entirely different quality if the percentage of orange is increased to eighty.

"Now analyze this room. How could light be used to define or articulate the space in this room? For example, walls define and enclose space. How could light be used in a similar way? How could color be used to more clearly define this room?

"Choose three qualities you would like this room to convey. How could you use color to achieve the effect of each of these qualities?"

"Color is life; for a world without colors appears to us as dead. Colors are primordial ideas, children of the aboriginal colorless darkness. As flame begets light, so light engenders colors. Colors are the children of light, and light is their mother. Light, that first phenomenon of the world, reveals to us the spirit and living soul of the world through colors."

Johannes Itten, *The Art of Color*.

preparation

several roofed, semi-circular or three-sided spaces, perhaps some of the cardboard spaces built in Creating Personal Space.

samples of textures previously made

hardware store spotlights or theatrical lights

colored gels

white butcher paper

white, black, red, blue, and yellow tissue paper

thumbtacks or tape

further exploration

"Analyze the restaurant spaces shown on page 17 and page 56. Which provides the greater sensory stimulation?

"At three different times on a sunny day, draw the pattern of shadows on the recessed windows or doors of a building facade. Study the modulation in form under various angles of illumination.

"Design a mural for an exterior wall of a clothing store, a food manufacturer or a warehouse using color to enhance the building's form. Draft a letter to sell the merchant on the value of buying your mural for his business."

check list

How many different moods was the participant able to demonstrate by manipulating the position and intensity of a light source? Was he able to articulate the effects of his manipulations?

Was the participant able to articulate the effect of the different color combinations that he created?

resources

Arnheim, Rudolph: *Toward a Psychology of Art*, University of California Press, Berkeley, Calif. (1966). Deals with the gestalt psychology of color perception and expression.

Bacon, Edmund: *Design of Cities*, Penguin Books, New York (1976). See Pages 53-57 and 242-251 on color as a dimension of progression through space.

Birren, Farber: "Color and Man-Made Environments — 1. The Significance of Light," *American Institute of Architects Journal* (Aug. 1972) 16-19. The biological implications of artificial environments.

Birren, Farber: "Color and Man-Made Environments — 2. Reactions of Body and Eye," *American Institute of Architects Journal* (Sept. 1972) 35-39. Biological, psychological, and psychic effects on man of the use of color and light in built environments.

Ehrenzweig, Anton: *The Hidden Order of Art*, University of California Press, Berkeley, Calif. (1967). Thoughts on color as affected by form.

Itten, Johannes: *Art of Color: The Subjective Experience and Objective Rationale of Color*, Reinhold, New York (1961). Students interested in further exploration of color will find Itten's work particularly valuable.

Ratliff, Floyd: "Contour and Contrast," *Scientific American* (June 1972) 91-101.

Strong, C. L.: "The Amateur Scientist," *Scientific American* (March 1971) 110-114. A method for generating visual illusions.

Wurman, Richard J.: "The Effects of Light on the Human Body, *Scientific American* (July 1975) 68.

Movement

Movement affects perception: exercises and field trips illustrate change in perception with movement and introduce time as it relates to the environment.

we live in motion

There was a young girl named Miss Bright,
Who could travel much faster than light,
She departed one day,
In an Einsteinian way,
And came back on the previous night.

Anon.

Historically, man's way of thinking about space was static, object oriented, and constant in time. This notion has been displaced by the twentieth century idea of the individual in dynamic, constantly changing interaction with his environment. In our culture, movement in and through the environment is the dominant manner in which we experience our environment. Moving in space and time is a way to perceive more wholly than is possible standing still.

As we move through it, we perceive our environment by sight; to a lesser degree by smell, hearing, and touch; and by the movement of our muscles and joints, which is "felt" by internal kinesthetic receptors. Our perception as we move is also colored by association based on past experience. We smell baking bread, hear bells, or see a pond; we are reminded of Grandmother, school, or swimming. These happy or unpleasant feelings partly determine our reaction to the new environment.

We experience movement by comparing our own velocity, size, and distance with that of the environment, which we perceive as being stationary. We visually interpret moving objects in the environment as rigid structures in relative motion. We ourselves are never completely static: our eyes, muscles, or other body parts are continuously in motion although we are often unaware of that motion.

Because our interaction with the environment is dynamic, any design, whether on the scale of a room, a neighborhood, or city, must be predicated on movement. Conversely, the design elements of any particular space define and limit one's movement through the space. Similarly, the placement of stationary elements, such as the windows in a building facade, can be used to control the movement of a viewer's eyes. For this to work as a successful design element, however, the velocity of the viewer must be taken into account. The faster we are moving, the less detail can be grasped, and therefore the larger the object must be to be comprehended; thus freeway signs are relatively larger than street signs.

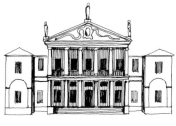

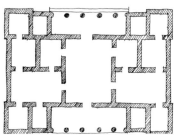

The static, bilateral symmetry of the residence *above* suggests the formal, stately motion of the 16th century. In contrast *below*, a dynamic balance of large and small elements expresses our fast-moving age.

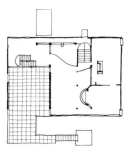
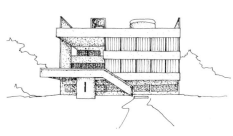

movement and perception

our experience
Participants should do this exercise one at a time so they are not distracted from reflecting on what happens to their perception as they move. Several tables can be set up if the group is large. Participants who have finished the exercise can act as recorder or instructor to those who follow. Or they can work on "movement is sometimes an illusion."

"Movement and our perception of movement are continuous forces in our daily living that have a subtle, but pronounced impact upon us. For instance, we stand still at an entrance, but our eyes sweep the room, recording an initial perception. Our impression of that room is quite different as we walk or run through it to a lounge chair at the other end. Our perception changes again as we sit in the chair and view the room from a new and lower angle.

"Think of other examples of your perception of movement. Recall the illusion of movement in stationary objects that often subtly induces us to move towards or past them. A round table pulls our eyes to follow its edge; a curved wall pulls us along beside it as we walk. Have you noticed that the illusion of movement alters as we move past the stationary object?"

Set up two platforms so that the top surface is at eye level for most participants. At the center of the first platform, place three identical articles in a cluster. Set up the second platform, with three identical articles of more complicated shape. Ask the participants to walk around the objects at an increasing velocity and articulate their perceptions to another participant acting as recorder for the group. Discuss their perceptions in terms of the concepts outlined in "we live in motion."

movement
in our time

"Traditionally, reality was perceived and thought of as static in design as well as in physics. Today, prevalent thinking considers reality to be dynamic and relative, an idea that is reflected in design. For instance, balanced asymmetry is often used instead of symmetry."

Walk with the participants toward a preselected site that encourages movement. The route to the site can also demonstrate many of the concepts involving movement. "As we walk to the site, point out examples of both symmetry and asymmetry at different scales. For example, a progression of different scales of symmetry can be seen as you perceive first the street pattern of high or low buildings, then the overall form of the building, next the placement of its windows, and last the fine-scale pattern or texture of the curtains at the windows.

"Identify design elements at different scales that suggest movement — a curving street, the ramp of a parking garage, a row of trees, small, slanted lettering on a signboard."

Arriving at the outdoor site: "Explore this space for five minutes. Make notes on how the space predisposes you to move through it."

Calling the group back together, remind them that this exercise involves their movement through the space. "How do your associations with this space cause you to move? For example, did your past experiences with buildings, creeks, trees, hills, influence the way you moved in this new space? What kinds of body movements did you use? Did you walk, run, swing your arms? Do you consider this a formal or an informal space? To what extent did this value judgment about the intended use of this space affect your behavior in it?

"How would you choreograph a dance to take place in this space? Compare the movements of the group in this space with the way in which space is filled, cut up, and defined by body movement in dance."

movement afoot

"Sketch the space you have just been in, retracing your movements, if necessary. Diagram your path through it. Mark the points at which you stood still. Compare your sketch with those of other participants. In what different ways have each of you manipulated this space by your body movements through it? Have you all 'cut' the space into the same shapes and volumes? What, besides 'conservation of energy,' determined your movements? How did the numbers, clusters, directions of movement, and noise generated by the rest of the group define and alter the space and determine your own movement through the space?

"Now we are going to look for illusions of movement. Examine the man-made features in the space, or nearby, such as paths, buildings, and pruned or sculpted vegetation. What decorative or design elements can you find that suggest movement? Do any of these elements 'force' your eyes to move in certain ways? Experiment by consciously trying to avoid the pull on your eyes to follow a curve or an arrow-like form.

"Identify static versus dynamic elements of design in this space. Which design elements have been used to support movement and which have been used to create a static, formal composition? Can you determine whether the more dynamic design elements were built more recently than the static elements?"

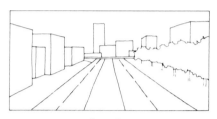

0 mph

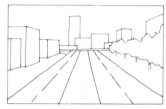

20 mph

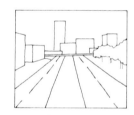

40 mph

60 mph

movement
as illusion

The forms of the freeway interchange, the stairs, the doorhandle, and the dome are architectural invitations to move our cars, our bodies, our hands, and our eyes.

"Architects throughout time have created illusions of movement. Can you recall any examples and explain how each illusion works?" (The Egyptians used repeating units to suggest a procession: parallel rows of sphinxes followed by columns, which together create a strong symmetry pulling you forward into the temple. Baroque architects, rejecting the straight lines of the Renaissance, used curves, which give their buildings a sense of exuberant motion both on the surfaces and in the interior and exterior spaces.) "A stairway is an architectural invitation to movement. Think of three or four stairways of completely different designs with which most of you are familiar. What does a stairway look like that makes you want to run up and down? that would encourage a stately procession? that would cause you to move very cautiously?

"Can you think of ways you can create an illusion of movement that is faster or slower than your actual speed?" (Compare a glass-front, with a closed elevator; compare an old highway with a modern freeway that is banked so as to minimize your sense of speed.)

a movie as an illusion

Prepare a sheet of thirty numbered squares that can be reproduced, cut, and made into a "man running" hand movie. The participants might draw stick figures for this purpose. Give each participant a "man running" sheet, asking him to cut out the numbered squares and put them together in order with a 2 ¼-inch spring clip. "Play with your movie and analyze what you observe. What happens? What control do you have over the movement? What is the scene at fast speeds?" (a man running) "at medium speeds" (a man dancing) "at slow speeds?" (different men posing). What does this tell you about perception?" (It is affected by velocity, that is, rate of change.) "When the velocity of the squares is very low do you still perceive movement? What sensations do you have when you run the squares backwards? Do you feel any disorientation or discomfort?" (A little discomfort is fun, such as in this exercise; a lot, is disturbing.)

perception during normal movement

Invite the participants in groups of three to walk a specified route and meet you at a chosen spot such as an ice cream parlor. When everyone has arrived, hand out a questionnaire asking, for example,

1　What kinds of functions were along the route?
2　Was there any visual homogeneity to the streets?
3　Was the route pleasant to travel? Why or why not?
4　How many city blocks did you walk?
5　Did the street width vary?
6　How many new buildings or construction sites were there?
7　How many curves were there in the street?
8　How many patches of open space or unbuilt land did you see?
9　What was taking place in the open spaces?
10　What items can you remember as being designed for car velocity rather than pedestrian velocity?
11　What was your approximate velocity along this route?
12　Draw a map of the route showing the above features as accurately as possible.

After the group has taken the same route back to the meeting room, have them again fill out the questionnaire.

"Pick a specific part of the route or the area around the ice cream parlor. What could you modify or add to the place to make it more ordered, memorable, functional, or esthetically pleasing?"

movement on wheels

Repeat this exercise along a similar route but moving at bicycle or car velocities. Compare the questionnaires filled out for each trip. "What different kinds of things did you notice at different speeds? Did the numbers of things you saw change? Did your perception of the distance you travelled change as the velocity changed? Does the map you drew for each velocity of travel contain the same types of features? If not, what are the differences? Do you think the routes you travelled work best for pedestrians, for bikers, or for drivers? How could each route be improved to make it work equally well for all three groups?"

our experience

There are several purposes to this exercise. Primarily, it stimulates the participant to notice more in his environment. Depending on the route chosen, the exercise emphasizes that too little order (chaos) is not memorable and is often disorienting. Conversely, it can emphasize the sterility and boredom of too little variety. (See Chapter 5, Order.)

65

Put together boxes or tables to create a surface which is eye-level when standing.

Obtain two sets of three identical geometrical objects at least eight inches in one dimension. One set should be of very simple form, such as triangular building blocks. The second set should be objects of slightly more complicated form.

Select an outdoor space which will encourage various kinds of movement in and through it, for example, a park with hills, a creek, and a playground.

Select a five or six block route to a convenient meeting place such as an ice cream shop.

Prepare a questionnaire to use in "perception during normal movement."

further exploration

"Analyze the movement patterns in your own residence. What traffic patterns exist for normal, day-to-day living? What traffic patterns do you find if your house is filled with a large crowd?

"Analyze the traffic patterns in public spaces such as in a shopping center.

"Spend some time standing next to a traffic lane of a freeway. Which senses make you most aware of the velocity of the traffic?

"Design a series of signs for a street which will be suitable for walking, driving, and speeding velocities of movement.

"Read the book by Lawrence Halprin listed below. On the basis of Halprin's ideas, try 'scoring' the group experiences of "perception during normal movement."

"Read the article by Gunnar Johansson listed below and repeat several of his visual perception experiments."

check list

Did the participant articulate the change in his perception of objects and their relationship to each other as his velocity increased? Was he able to relate this experience to a large-scale environment such as a room or a neighborhood?

Was the participant successful in creating the illusion of movement? Was he able to explain how and why it worked?

resources

Arnheim, Rudolf: *Art and Visual Perception, a Psychology of the Creative Eye*, University of California Press, Berkeley, California (1974). Chapter 8 is a lucid account of movement and perception.

Environmental Psychology: Man and His Physical Setting, H. Proshansky, W. Ittelson, and L. Rivlin (editors), Holt, Rinehart and Winston, New York (1970). The papers by Thiel (Pages 593-619) and Bechtel (Pages 642-645) describe methods of studying movement in space.

Halprin, Lawrence: *The RSVP Cycles*, Braziller, New York (1970). Creative processes in the human environment.

Johansson, Gunnar: "Visual Motion Perception," *Scientific American* (June 1975) 76-88. An interesting theory of why moving stimuli are not perceived as blurs.

Leonard, Michael: "Humanizing Space," *Progressive Architecture*, (April 1969) 128-133. A discussion of movement as determined by the shape of space.

Scale

10

Perceiving size: design problems illustrate how scale can be used to enhance function and to provide orientation, enrichment, and symbolic communication in our environment.

how big does it seem?

scale and proportion

We are comfortable in our relationship to objects when they are in scale with us.

We perceive with ourselves as our own reference point. *Scale* is the perception of the size of an object relative to ourselves. *Proportion* differs from scale in that it denotes the relationship two objects have with each other or the relationship an object has with its context.

Perception of scale involves both close and distant sensory reception. In our society, scale is perceived principally through our vision, which interprets for the other senses.

Scale and proportion act as measuring tools for personal orientation in the environment (time and space). A water tower orients us in space because it is relatively taller than surrounding buildings. The scale of a city changes as the city grows over a period of time. Thus the eye is able to register a difference between "now" and "then."

Scale can be used to communicate symbolic messages in a particular context; for example, a throne communicates power because it is bigger and higher relative to other chairs.

Scale and proportion can be used to help order the environment. By building all the homes in a neighborhood of similar size, a constant scale is maintained and a sense of order introduced. Enrichment and variation in the environment can also be achieved by careful use of different scales. A windmill or silo provides visual relief from the unending sameness of the prairie.

Contemporary forms in the scale of the original structure are used successfully to blend a new addition with an old building in Assisi, Italy.

Man's perception of scale has periodically altered. The variables that cause the differences between one era's sense of appropriate scale and another era's are (1) man's view of himself in relationship with time and space and (2) the technology he has available to transform the ideas into physical elements such as the Eiffel Tower in the nineteenth century and the moon lander in the twentieth.

Our age is concerned with appropriate scale or proportion for a much wider range of activities than heretofore. In our mobile society appropriate scale for individuals in motion at different speeds is indispensible. It affects safety as well as efficient functioning and quality of experience.

scale, a tool for orientation

Begin a discussion of the concepts "scale" and "proportion." Record key words and phrases. For convenience, we have chosen the following as our operating definitions of the two words:

Scale denotes the psychological perception of size an individual senses as his relationship to an object or space.

Proportion is the relationship two objects have with each other or the relationship an object has with its context.

The water tower is an orientation point very much in proportion to the city. It is greatly out of proportion to its neighbors.

If you lived in this house would you feel sheltered or overwhelmed by the tower?

entrances

Discuss the idea that different entrances are designed differently for various functions. One of the differences can be the size of the door in relation to human size. This use of scale can be used to transmit a symbolic message: the bigger the door the bigger the person. For instance, the Pope, King, or Emperor uses a massive entrance that reinforces his exalted rank. Often encased in one of the huge ceremonial doors is a door small enough to be physically humbling for the common man. Prior associations register the doors as symbolic of the relative status of the users.

Other entrances could be those appropriate for (1) a small cozy flat, (2) a hall for the Queen of England, (3) a boutique shop, (4) a room for the President of the United States, (5) an amusement park, (6) a supermarket. Or consider a park entrance that will permit walkers, bikers, horseback riders, and automobiles to enter without conflicting with one another. "What kinds of entrances can you think of? In your examples, where does the entrance begin? at the threshold? at the sidewalk? at the street?"

Ask the participants to work as individuals or in teams and assemble butcher paper or cardboard sheets into surfaces at least ten to fifteen feet high by eight to ten feet wide.

"We are going to design entrances that primarily use *scale* to communicate a symbolic message. Choose a functional and symbolic intent to convey in an entrance. Write it on a signed card and hand it in. Now, design your entrance only in terms of scale and a basic shape. Leave details of decoration until later. Begin by drawing a scale figure your own height. If you prefer to work at a smaller scale, use a scale figure at least three feet high."

What symbolic message does each of these entrances communicate to you?

After the teams have finished their basic designs, suggest "To further reinforce your message, use other design elements such as texture, color, line, pattern, rhythm, balance, symmetry, and proportion on your entrances." Ask the participants to select two entrances that seem to work particularly well, but are entirely different. Discuss and compare them.

"What messages do the scale and basic shape convey?" (Compare with the designer's intended messages.) "What other elements has the designer used to amplify his message? Do any of the elements seem inappropriate in relation to the basic messages? What associations do you have with the entrances?" Discuss each entrance in terms of both scale and proportion.

Now ask each participant to visit the other entrances designed by the group. Ask the participants to list the adjectives they associate with each entrance, leaving a card so that the designer may compare it with his original design intentions.

our experience
Help each participant execute his original design and discourage his changing to another intent. Emphasize the decision-making process rather than the final product. It is helpful to work first in pencil.

We have found that much is learned while discussing well-designed entrances.

scale and proportion in the environment

Ask the group as individuals or together to choose a street with buildings of varying heights and set-backs (perhaps in a transitional neighborhood). "Which buildings or objects — for example, billboards — appear to be out of scale with you or out of proportion with the other buildings on the street?

"Analyze the scale of the street in terms of your rate of movement. What size objects (house, windows, door knobs) do you easily perceive from a car at thirty miles per hour? on a bike? while walking? How do the problems of scale and velocity relate to the design of signs along the street? How has scale been used symbolically along the street?" (Billboards symbolize commerce; rows of large oaks symbolize leisure; a divided street with concrete median symbolizes speed.)

visitors from a small planet

Give butcher paper to the participants and say something like "Design this room so that it will function for our guests from Mars exactly as it has been functioning for us. Design it so that they can use it with the same degree of comfort as we use it.

"Our problem is twofold: (1) our guests are like us in every respect except one — they are two feet tall; and (2) the room must be designed so that both we and the Martians can use it at the same time."

Discuss the participants' solutions, what they did, and why they did it that way. "How did you arrive at your solution? Did you visualize yourselves as Martians? What activities did you consider yourselves as sharing with the Martians? Or did you create a 'small' environment within the large one? What changes would need to be made in *this* building to accommodate the Martians?" (stairs, water fountains, door handles . . .)

preparation

white butcher paper

duct tape

masking tape

cardboard (reused from Creating
Personal Space)

straight edge

broad-line felt-tipped pens

scissors and matte knives

ladder or some way to hang paper
from the ceiling

further exploration

"Illustrate how scale and proportion are used annually by the fashion industry to outmode your wardrobe.

"Put together a slide show to illustrate how scale can be used symbolically."

check list

Can the participant articulate the effects on himself of the use of scale in a particular situation?

Can the participant articulate what he considers is the appropriate scale for various situations?

Is the participant able to manipulate scale to achieve a predetermined effect?

resources

Asimov, Isaac: *Fantastic Voyage*, Bantam Books, New York (1966). A trip through one's own body.

Bacon, Edmund: *Design of Cities*, Penguin Books, New York (1976). See Pages 217 and following for the changing scale and proportion of cities looked at historically.

Carroll, Lewis (Charles L. Dodgson): *Alice's Adventure in Wonderland.* The first two chapters describe Alice's growing and shrinking as she goes through the rabbit hole and enters Wonderland.

Fast, Howard: "The Vision of Milty Boil," and "A Matter of Size," *Time and the Riddle*, Ward Ritchie Press, Pasadena, Calif. (1975). Two short stories that question our assumptions about human scale.

Scholfield, P. H.: *The Theory of Proportion in Architecture*, Cambridge University Press, London (1958). A detailed analysis of how visual proportion is employed in classic, renaissance, revivalist, and contemporary architecture.

Symbols

11

Reading the symbols around us: recognition of symbols as communication tools leads to the ability to analyze the symbolic content of the environment.

symbols are shortcuts to essential information

Symbols, both verbal and nonverbal, are means of communicating; they carry associative messages. A symbol may evoke a highly emotional and personal response, and thus call forth many associative patterns from many levels. Associative patterns are learned on an individual level, on a family level, and on a cultural level. Because associations are so abundant, a symbol is a shortcut to essential information.

Symbols may have highly specific meanings to individuals, families, or cultures, or they may have universal meanings. A specific symbol has a particular meaning to a particular individual, depending on the history of his experiences (associations) and the context of that symbol.

Each individual's self-image is in part defined and enhanced by his personal relationship with his heritage and with the universal values of his historic culture. Symbols are a means by which he relates himself to, and orients himself in, time and space.

Change characterizes our lives. Certain physical symbols that remain constant can be a way of fixing time, of making it stand still. A symbol may also be a psychological tool for our orientation by conveying to us a sense of our roots.

The nonverbal symbols significant to every culture are both man-made and natural. Many sounds, smells, forms, textures, and body movements have significant symbolic meaning. Symbolism is a language to be used.

This horse and rider and the dome behind are symbolic in many ways. The dome can be seen for miles away and identifies the seat of the state government. The Capitol with the Goddess of Liberty on top connotes continuing democratic government.

The cowboy links us with a bygone age; he symbolizes the bravado and rugged independence highly valued by our society. Cast in bronze, he projects a sense of permanency that will withstand a change. The statue may also evoke personal associations for an adult who climbed upon it as a child.

Lost in history is the sculptress, Constance Warren, whose father's western stories inspired her work. Her symbol survives her memory.

Dolls are used by children to symbolize characters in the adult world around them. For adults, dolls symbolize their past.

exploring symbolic messages

"Think of that place you consider home. Imagine your feelings as you see it burning down. What would you try to save? Is the value of that object you save intrinsic, or symbolic? What sentiment is bound up in it? Why? Do personal symbols have the most intense meaning for us? What does 'home' symbolize? Why do we have a strong feeling relationship with it?"
(Remember that the home is a physical representation of one's identity.)

"What is a symbol? How do you use a symbol? Why do you use a symbol? What kinds of symbols can you think of? What are some of the physical symbols of the town or city you grew up in to which you relate personally? Are there any natural symbols of your town or its region to which you personally relate?" Assemble into teams of four or five persons. Ask each team to study the categories presented in the text, discuss the examples used, and select additional examples of each to present to the entire group. "Emphasis should be on symbols in this community. What categories and examples can be added to those in the text?"

mementoes

"Of what value, aside from occasional use, are the old family Bible or family chest, both often stored in the attic? What do they symbolize to you? What are some man-made objects that, through your ancestors, connect you in time?" (antique furniture, jewelry) "connect you with history?" (family records or diaries) "Which of your family's mementoes have the most meaning to your mother? to you? Are they the same ones? What are some natural symbols that connect you and your family in time? Is there a tree you climbed as a child? Or a stream near your childhood home?" (Perhaps there is a family vacation place identified with distinct natural characteristics of a particular region.) "Do you choose the same kind of Christmas tree that your family chose?"

our experience
The pictures or slides suggested under *preparation* can be used to help define symbols and to start the group discussion.

Adult groups very quickly grasp the nature and use of symbols. Younger students take much longer. If you don't get much reaction in the first discussion, ask the participants to imagine the same problem assuming they are each thirty-five.

For young students, a follow-up activity to these discussions is almost indispensable to help coalesce the variety of concepts.

74

power

"How can you tell that someone has power? Part of having power is making others feel powerless or afraid. Symbols are the medium used to express the message 'power.' Greater control is obtained if the power is implied and does not have to be exerted directly. What symbols of power are worn?" (badge, hat with braid) "How are objects used to symbolize power?" (the king sits on a throne; the President rides in a limousine) "Does the same object symbolize power in every culture? Would an umbrella, which symbolizes the authority of a Shinto priest in Japan, be considered a symbol of power in any other culture?

"How is space used symbolically to connote power?" (a space, relative to the spaces around it, is made larger, has more windows, is appointed with materials that symbolize luxury) "Describe the space occupied by the most powerful person you know."

give me your tired, your poor...

"What are some symbols of our society?" (There will be many mentioned. Eventually someone will mention the flag, a good symbol with which to emphasize the importance of symbols to individuals and society and why they should be preserved.) "What does the flag mean to you? On how many levels do you relate to it? Of what use is such a symbol to a person? to a society? In what ways does it help provide identity in time and space? Does the flag have different meanings in different contexts? Look at the slides or photographs of the flag in different contexts and consider how the context affects the message." (For example, a flag flown upside down is a more or less standard symbol of distress. A flag sewn on the seat of bluejeans was a symbol of protest in the sixties.)

eating out is fun

"Eating out is one of the activities we engage in as groups. There are many social activities that must take place *somewhere*, such as in a restaurant. What other activities can you think of that are symbolic of our culture and must have a place or 'container' in which to be carried out?" (sports, movies, parades) "A historic example would be the Roman coliseum.

"What are some symbolic activities that you associate with other cultures? Do the places themselves, as well as the social rituals, then take on a symbolic quality? How is space used symbolically?" (big office for the boss; windowless cubicle for his secretary — symbolic both of their relative positions and of the boss putting someone between himself and the public) "What one space most symbolizes your city to tourists? to the residents?"

gateway to the west

"An arch was chosen to symbolize St. Louis's position at the beginning of the West. What other man-made symbols represent to us today the historic struggle of the plainsman with his physical world?" (barbed wire fences, silos, courthouses) "What symbols of American Indian culture have been brought into modern United States use?

"What are some natural symbols we associate with the plains regions?" (expanses of grass, scrub trees, 'blue' northers) "What are some built and natural symbols of other regions in the United States? What does Plymouth Rock symbolize? Would it matter if Plymouth Rock is not the original rock on which the pilgrims landed? What natural symbols do we associate with the old south? How do we use Hispanic symbols?"

symbol of india and...

"An object may symbolize more than one thing. Shah Jehan built a tomb for his beloved that is a universal symbol of romance. A 'Taj Mahal' is also a symbol of timeless perfection. Used derisively, its name may imply too expensive a building. To a historian, the Taj Mahal symbolizes the fusion of Persian, Islamic, and Indian cultures brought about under the Mogul dynasties. What do the Parthenon, Chartres Cathedral, and the Lincoln Memorial mean to you? Those monuments are also cultural symbols.

"What are some symbols of technological advancements?" (pyramids, Eiffel Tower, airplane, automobile, space capsule) "What are some things in nature that symbolize a continuity with our past?" (redwood trees, dinosaur tracks, the great forests of Siberia)

thumbs down

"Gestures are a universal form of communication. What other body language carries a symbolic message in our culture?" (eyes: raised eyebrows, rolling eyes; arms: stiff arm salute, folded arms)

"What sounds carry a symbolic message in our culture?" (sirens, bells, chimes, whistles, jack-hammer) "Relate this to the built environment. Why is it so enriching to have sounds in the environment? What tastes or foods carry a symbolic message?" (peaches: peachy; lemon: clean, fresh; Pablum: dull, mushy, spineless) "What smells carry associative messages?" (carnival foods, greenhouses, the artificial scents used to create an atmosphere)

where am i?

"This building is a sign that identifies the location of Best Products; it is also a symbol, but of what? What is the difference between a sign and a symbol? What other symbols are used by businesses for identification? What are some graphic symbols used to communicate directions? Unambiguous graphic design is necessary to communicate information clearly and safely. Is the particular symbol memorable? How well does it associate with the intended message? 'This way to the . . .' does not need the word 'circus' to indicate what the sign leads to if it is painted in balloon letters. What lettering would you select to combine with a graphic symbol to identify a funeral home? a space colony? yourself?"

presentation

Ask each team to present their new symbols and categories to the group. Which team had the greatest number of symbols not used by others?

symbols of the past
Old-timers see in this photomural a personal link with the past, a symbol of the trains that no longer arrive and depart from the station that stood on this site. To newcomers the mural is just an unexpected decorative wall.

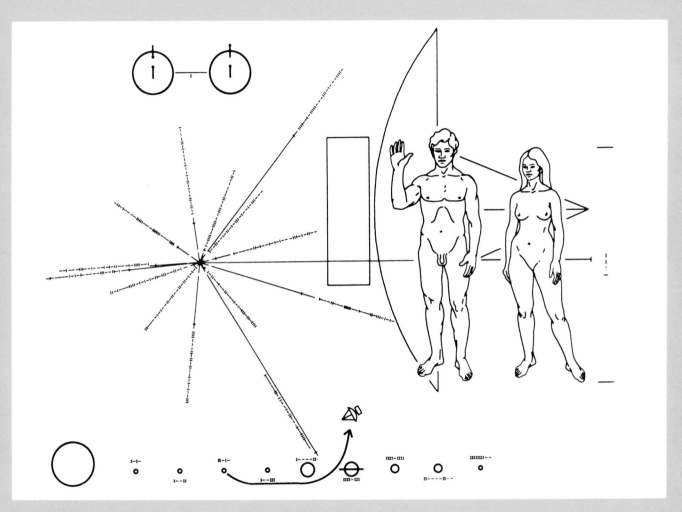

How do you communicate to an
unknown society outside our
galaxy with a message that will
not be intercepted for at least
80,000 years? Pioneer 10, the first
spacecraft to leave our solar
system, carries a message about our
civilization etched in symbols on a
gold-anodized aluminum plate.
What do the symbols
communicate to you?

preparation

35mm slides or pictures of the flag in various contexts

pictures of road signs and other designed symbols of less obvious meanings

(The pictures and slides can be used to help define symbols and to start the group discussions.)

Assign each person to bring three objects having symbolic content as preparation for Designed Objects, Chapter Twelve.

further exploration

"Design in any medium a symbol for school, this group, or the city. Display it at the next meeting with a concealed statement of what you have tried to symbolize." Ask each participant to write down the meaning of this symbol to him, and compare it with the intended message.

"Draw from memory and analyze a group of traffic signs. In how many ways is each message expressed? Can an illiterate 'read' the message?

"Work out a slide show illustrating various kinds of symbols.

"Write a history of the peace sign which includes its acceptance or rejection by various groups."

check list

Can the participant give examples not mentioned in group discussions for each category of symbols?

Can the participant analyze the symbolism that various objects, people, and places hold for him?

resources

Dreyfuss, Henry: *Symbol Sourcebook*, McGraw-Hill, New York (1972). An attempt to design graphic symbols for all facets of life.

Fast, Julius: *Body Language*, Pocket Books, New York (1971). The use of behavioral symbols.

Jencks, Charles: *The Language of Post-Modern Architecture*, Rizzoli International Publications, New York (1977). An entertaining survey of symbolism in architecture.

McLuhan, Marshall: *Understanding Media*, McGraw-Hill, New York (1964). Changing means of communication and their affects upon us, including an analysis of advertising in our culture.

Venturi, R., Brown, D. S., and Izenour, S.: *Learning From Las Vegas*, The MIT Press, Cambridge, Mass. (1977). A fresh insight into what constitutes "architecture."

Designed objects

12

Perceiving wholes and analyzing components: learning to articulate personal responses to an object initiates evaluation of all components as they contribute to the whole object.

design is an ordered relationship of components

The process of designing is a way we organize sensory data into meaningful wholes. Any designed object or environment may be thought of as a whole composition to which form, function, texture, line, color, scale, process, and symbolic content contribute as components. When we line a road evenly with trees, we make functional and esthetic design. We also create a design when we paint trees on a canvas to hang on our wall.

A design will be deeply satisfying if the beholder can respond from the perspective of his own culture to the direct qualities of the design. The tea ceremony in a meticulously designed room is ultimately satisfying to the Japanese who has trained his body and mind to the discipline of following a rigidly prescribed pattern. To a Westerner, it can be a muscle-straining, incomprehensible ceremony that he can appreciate only because to him it is exotic.

When we experience an object we instantly register a first impression. Then, usually subconsciously, we begin simultaneously to synthesize and to analyze its parts. The pitcher feels rough, is a pretty color, pours well. Conscious perception of the whole is limited by our ability to perceive numbers of elements simultaneously. Thus, on a conscious level, our response may be simply "I like it" or "I don't like it."

Often, personal associations from past experience can have an overwhelming influence on our value judgments. We take an instant dislike to a teacup solely because its decoration triggers associations of having to clean our plates down to the flowers at Grandmother's house years ago. A conscious or unconscious analysis of how well each of the components of an object is executed and of how well it contributes to the whole is another way by which we form personal value judgments.

When a group in a particular culture agrees upon a value judgment and declares an object to have intrinsic "esthetic excellence," the influence of personal associations is minimized. A connoisseur of teacups sees beyond his personal associations to appreciate the form, texture, quality of glaze, color, method of manufacture, and decoration.

In our society, it is the artists who explore new forms. Some of those forms become the new norms; some are passing fads. A literary work of art does not become accepted as a great book for several generations. Books that do not "make it" disappear into oblivion. However, china, chairs, and houses that are not great remain in use. They do not have to be intrinsically excellent to survive. Because they do survive, they become a part of our past associations and experience, influencing our personal perception of new forms.

The two design motifs illustrated at the left became popular fads soon after each was introduced in the 16th century and in the 1940's. Today, both are considered classics of design. Do we like them now because each just happened to survive, or because each has an intrinsic excellence?

analyzing objects

Display the objects for the group to view. List the following words and elicit definitions of each: texture, color, mass, process, scale, form, structure, context, and function. Divide into teams of three or four participants and give each team an assortment of objects and the questions below. Suggest that one way to begin the discussion is to group the objects into those everyone likes, those everyone dislikes, and those on which there is no agreement. Each object can then be referred to, in turn.

1   "What is your response to the WHOLE object? If you like it, do you like everything about it, or can you suggest improvements? If you dislike it, can you analyze why? Are you rejecting the object as a whole because of disliking one or two of its qualities? Does the object have a 'farout' design. If so, what makes it different; that is, 'farout'?

2   "In what CONTEXT can you visualize this object? Can you imagine the object being used by anyone or only by certain categories of people? Would you like the object in any context? Is there any category in which you would dislike it?

3   "What response do you have to the FORM of this piece? What associations do you have with the shape? Pleasing? Unpleasant? Why? Does the form connote one style or historical period? Is it in fashion now?

4   "What STRUCTURE of the object makes it stand up or holds it together?

5   "How do you feel about the COLOR or combination of colors? Why? How does the color work with the shape and the texture?

What makes a favorite chair special?
How do the
shape
size
scale
texture
color
structure and
manufacturing process
contribute to the design of each of these chairs? What is the intended function of each chair? Would the function change if the context of the chair were different?

The chair
A funny thing about a Chair:
You hardly ever think it's *there*.
To know a Chair is really it,
You sometimes have to go and sit.
           Theodore Roethke

6   "What response do you have to the TEXTURE or combination of textures? Why? Does the texture fit the function of the object?

7   "What feeling of MASS does the object have? Is it complementary to the function of the object?" (A tall vase should be heavy, but a pitcher must be light enough to lift.)

8   "Is the SCALE of the object appropriate to its function?" (tiny cups for espresso, large cups for ordinary coffee) "How does the scale work with the design?" (If it is a small object, is its texture also small?)

9   "What does the object SYMBOLIZE to you? Does it have a historic, a cultural, or a personal symbolism which explains your response to it?

10   "What do you know about the PROCESS by which this object was made? Was it extruded, cast, fabricated, or molded? Was it made by machine or by hand? Does it make a difference? Why?

11   "What is the FUNCTION of the object? Does it have more than one function? Compare two objects that have the same function. Which do you think works the best? Do you all agree? How do the form, structure, texture, color, mass, and manufacturing process affect the function. Do all of the components of the object complement one another, or do some of them contradict?

12   "Has the group consensus about any of the objects changed as a result of this analysis? What about your personal judgment? Are there any of these objects that you personally dislike, but that you can agree with the group have an intrinsic excellence?"

preparation
Collect objects exhibiting variety in shape, texture, color, and function. Some objects should have historic value. Include objects of both good and poor esthetic quality. Cups, bowls, pitchers, and vases are particularly suitable.

Encourage the participants to bring items of personal symbolic value.

further exploration

"Respond to and analyze, as a composition of objects, this meeting room or the room in which you sleep." A written analysis will help crystallize the concepts of this session and provide a good base for the following, much more loosely organized, Designed Environments field trip.

check list

Is the participant able to perceive and discuss the various components of an object, their effects on him, and why they affect him as they do?

Can the participant articulate how the components work together to form the composition and how the composition relates to the functions of the object?

Is the participant able to articulate his evaluation of an object and explain how and why he has drawn his conclusions?

resources

Downer, Marion: *Discovering Design*, Lothrop, Lee and Shepard, West Caldwell, N.J. (1947). Photographs evoke awareness of hue, pattern, and rhythm in nature and in man-made objects.

Gowans, Alan: *Images of American Living: Four Centuries of Architecture and Furniture as Cultural Expression*, J.P. Lippincott, New York (1964).

Masters of Modern Art, A. H. Barr, Jr. (editor), The Museum of Modern Art, New York (1958). Distributed by Doubleday. Includes designed objects in the collection of the Museum of Modern Art.

Norberg-Schulz, Christian: *Existence, Space, and Architecture*, Praeger, New York (1971). See Pages 31-32 on "things."

Norberg-Schulz, Christian: *Intentions in Architecture*, The MIT Press, Cambridge, Massachusetts (1965). See Pages 64-66 on how we orient ourselves in three basically different ways to objects.

Designed environments

13

Creating an environment by relating parts to a whole: a field trip is taken to compare environments having different qualities but the same function.

environments
are compositions

Each component of this office space contributes in more than one way to form a whole. The chairs move for audio-video presentations or conference groupings; the upper railing is also a light source; the cylinder with the decorative arrow encloses the stair.

The ceiling
Suppose the Ceiling went Outside
And then caught Cold and Up and Died?
The only Thing we'd have for Proof
That he was Gone, would be the Roof;
I think it would be Most Revealing
To find out how the Ceiling's Feeling.
           Theodore Roethke

To design or evaluate a small environment such as a living room or a church is more difficult than to analyze an object, but the process is the same as that experienced in previous chapters. Environments may also be thought of as compositions made up of parts. In these small-scale environments, the parts are the furniture, the windows, the landscaping. The parts in themselves have qualities such as form, color, function. Together, the parts make up the form, color, and function of the environment. In addition, as the scale increases from teacup size to room size to city size, the relative importance assigned to specific qualities of the parts changes. A teacup will be judged primarily by its color and shape; a city will be evaluated as to how well it functions. When the components are well designed and successfully combined, an object or an environment can be judged as "good" or "excellent."

"Excellent," "successful," "well designed" are value judgments much more difficult to make about even a small environment than about a teacup. The individual must respond synthetically to a much larger, whole experience and analyze a much greater number of parts and their components in terms of more complex criteria. We walk into a strange room and perceive a total impression of the room. Simultaneously, we analyze its components — the fabric on the chair, the height of the ceiling, the patterns of windows, light, floors.

In evaluating how well an environment functions, it must be judged for those qualities that aid or detract from living processes. A particular environment must, for example, be safe, provide shelter, and be arranged to facilitate the activities carried out within it. It must also provide satisfaction — pleasure, rest, excitement — for its users.

a trip to small-scale environments

At each preselected site, divide into small groups. No two groups should be in the same space at the same time. Stagger the route through the spaces and allow about fifteen to twenty minutes per area. Ask the group to quietly observe the characteristics of each space and how it is used by the people in it. After this period of silent observation, the group can be asked to respond to leading questions such as these:

1   "If you had to choose one word to describe this space, what would it be?" (There are no "right" answers.)

our experience
We have found that controlling how these spaces are experienced makes considerable difference in how they are perceived. People are too accustomed to depending on what the other person thinks.

This is a time to start recognizing the process of making personal value judgments. Some participants can incorporate your suggestions without feeling that the decision has been taken away from them. To them you can introduce factors that might lead to a different judgment. Others, just learning to *express* personal judgments need to be left alone for the time being. It is sometimes necessary for you to encourage participants to commit themselves, however temporarily, to a judgment. *Any* judgment, however, must be accepted by the instructor as a valid *personal* opinion whether or not it expresses "good taste" as defined by a recognized design critic.

In the context of this course or workshop, you the instructor, reflecting your membership in a particular subculture, are the one who ultimately defines quality. If you want the participants to learn your criteria for quality, then say so honestly and clearly. If you want them to define their own standards of quality, fine. But in the latter case, you should proceed with the understanding that in helping them, there is no way (nor is it necessarily desirable) to eliminate your own viewpoint.

These are two spaces in which to eat within the same dwelling. What are the components that give each place its individual character? What other functions can you visualize taking place in each?

2 "What are the components of design used in this space? How are the components used to define order?" (structure, color, texture, light) "Is it easy to orient yourself in this space? Is the ordering too rigid to interest you? Which components made the greatest impression on you? Why? Are there any other components that impressed you in some lesser degree? How and why? How do these elements limit, expand, or complement one another?

3 "How would you evaluate the over-all quality of this space? If you could alter the quality, what would you do? Can the space be put to a variety of uses? What component or combination of components works best for the intended function of the space?

4 "What assumptions might you make about the personalities of the people who occupy this environment? Explain your assumptions."

Proceed to the second site. Take time to compare the two environments.

5 "How do the two environments differ and why? In which space would you rather spend the greater amount of time? Why? Both of these environments have been planned to serve similar needs. Which functions better to meet these needs?"

further exploration

"Write an essay in which you recall the effect of specific qualities of the visited environment in terms of the elements of space: volume, proportion of doors and windows, real and artificial light, materials, colors, structure, objects, scale, and texture. Include the relationship of the inside environment to the outside and of the whole building to its physical context."

check list

Did the participant relate his experience in these real environments to the previous exercises?

Did the participant articulate both a synthetic and an analytic response to the quality and appropriateness of the components within a total environment?

resources

Baldwin, James: *Go Tell it on the Mountain*, Dial, New York (1963) 39-40. Yglesias, Helen: *Family Feeling*, Dial, New York (1976) 138. Two very different reactions to the Forty-Second Street New York Public Library.

Clark, Kenneth: *Ruskin Today*, Holt, Rinehart and Winston, New York (1964). See particularly the nineteenth century critic's writings on architecture, Chapter 4.

Cooper, Clare: "The House as Symbol of the Self," *Designing for Human Behavior*, J. Lang et al. (editors), Dowden, Hutchinson and Ross, Stroudsburg, Pa. (1974). The theories of Carl Jung applied to the understanding of why our "house" has special significance to us.

Rasmussen, Steen Eiler: *Experiencing Architecture*, The MIT Press, Cambridge, Mass. (1962). A lucid introduction to the excitement of architecture.

Venturi, Robert: *Complexity and Contradiction in Architecture*, Museum of Modern Art, New York (1967). A provocative argument that architecture should express the ambiguity and richness of our modern experience.

Zevi, Bruno: *Architecture as Space: How To Look at Architecture*, Horizon, New York (1957). An analysis suggesting that the essence of a building is the space it encloses.

Photointerpretation

14

Analyzing built environments: participants photograph and discuss the relationship of components of the built environment.

we must learn not only what to see, but how to see

Seeing with the camera is another way to become more aware of the environment. The decision to take a picture is predicated on observation, judgment, and evaluation. The fact that the decision is recorded, and thus acted upon, indicates some degree of willingness to stand responsible for one's decisions about environment. People are generally unwilling to take responsibility for their judgments. Perhaps this unwillingness stems from fear of making the "wrong" decision or fear of ridicule from one's peers. Whatever the inhibitor, it often leads to action based not on one's own convictions, but on the anonymous consensus of the group.

Seeing with the camera also demonstrates the difficulty of communicating about the environment. A drawing or photograph is a two-dimensional communication about a subject that is four-dimensional if one includes the effects of time. The only stimulus is visual.

Discussing one's own photographs usually causes one to see in a picture much that was not perceived when it was taken and sharpens the ability to articulate what is perceived. If the picture is discussed soon after it is taken, the photographer is often able to describe elements that make up the whole — elements that a camera cannot record, such as sounds, odors, the weather, and how the place in the photograph is used.

The environment cannot be completely pictured, nor can it be wholly described. Verbal communication is limited to language, a linear process in time. Language is also used symbolically, which constricts our ability to comprehend and communicate experiences taking place in the environment. Nevertheless, learning to communicate one's decisions and the rationale on which they are based can be helpful in learning to think both analytically and synthetically. It is part of learning the decision-making process.

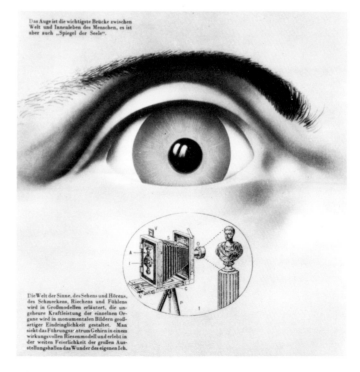

Das Auge ist die wichtigste Brücke zwischen Welt und Innenleben des Menschen, es ist aber auch „Spiegel der Seele".

Die Welt der Sinne, des Sehens und Hörens, des Schmeckens, Riechens und Fühlens wird in Großmodellen erläutert, die ungeheure Kraftleistung der einzelnen Organe wird in monumentalen Bildern großartiger Eindringlichkeit gestaltet. Man sieht das Führungszentrum Gehirn in einem wirkungsvollen Riesenmodell und erlebt in der weiten Feierlichkeit der großen Ausstellungshallen das Wunder des eigenen Ich.

"The eye is the most important bridge between the world and the inner life of man, but it is also the 'mirror of the soul.' "
Herbert Bayer
The Function of the Eye, 1935

photostudy of design components

Divide the teams of two or three participants, each with a camera, two rolls of film, and a copy of the worksheet for a field trip. Demonstrate how to load and use the camera. "You should include in each photograph as many items on the worksheet as possible. You will be asked when you return to explain why the picture you took illustrates the stated points. This is an opportunity for you to demonstrate how well you have translated into working tools, all of the ideas we have discussed."

space

1   Record a small open space whose *shape* is defined and pleasing to use.
2   Record an open space that is poorly defined and discourages use.

process and function

1   Record an environment that is undergoing change. Record one or more steps in this process of change.
2   Record a built environment in which a lack of successful planning is obvious and renders the environment difficult to use.
3   Record a physical arrangement for organizing people into a moving line. Indicate whether or not it is successful on the basis of all the criteria we have studied.
4   Record a place that works well for an individual to sit safely, rest, and contemplate.

form and structure

1   Record, from the outside, two built structures whose forms are pleasing to you and are functional.
2   Record two cantilevered structures.
3   Record two structures built from prefabricated factory components.

pattern

1   Record two instances of order achieved by visual rhythm.
2   Record three different kinds of patterns of natural materials.

texture

1   Record a man-made texture that you like.
2   Record a natural texture that you like.
3   Record a man-made composition that uses several textures together successfully.
4   Record a man-made composition that uses several textures together unsuccessfully.

light and color

1   Record a use of light that has changed a place from one that appears unsafe to one that appears safe.
2   Record an example of how light changes the apparent form of a building.
3   Record the successful use of a color on a built structure.
4   Record a built structure in which several colors conflict with or confuse the intended function or message of the structure.
5   Record a building whose form has been significantly altered by the use of color.

movement

1   Record a built environment through which pedestrian movement is fun and is encouraged by the design.
2   Record a built environment through which pedestrian movement is awkward and discouraged by the design.

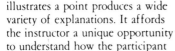

our experience
There are no "right" answers in this exercise. We have found that discussing why a particular picture illustrates a point produces a wide variety of explanations. It affords the instructor a unique opportunity to understand how the participant perceives the course and its contents and how he assimilates what he has selected to "learn" to this stage.

We preceded the field trip with a review of the first thirteen chapters and the worksheets.

Working in the field in groups of two or three enhances the learning process.

If the exercise is to be used as a test, the group should be notified in advance to forestall their hostility.

3   Record a conflict of automobile and pedestrian movement.
4   Record a situation in which pedestrian and automobile movement have been made compatible.

scale

1   Record two examples in which the designed scale of buildings or components is wrong for the intended function.
2   Record an environment intended to "humble" a person.

symbols

1   Record a "double message"; that is, an environment whose literal message is contradicted by a symbolic message.
2   Record a building in which the literal message is reinforced by the symbolic message.

analysis

Form discussion groups composed of three to six of the field-trip teams. Ask each team, in turn, to present their photographs to the group for evaluation. The rest of the group should ask as penetrating questions as possible to force the team to justify their choice of examples.

analysis: interviews

The instructor may interview each participant individually about his photographs, hearing him justify his decisions and probing with him to determine how well he has grasped the ideas of the course. Because each participant's perceptions are unique, his responses will arise from his personal context. As the discussion progresses, the instructor can help the participant to find the larger, more impersonal context, and thus gain significantly more from the interview.

an instant camera and two packs of
film for each two participants

a copy for each person of
worksheet "photostudy of design
components"

further exploration

"Describe with words and photographs your workplace. Analyze in detail its
faults and its virtues in the total context of your getting to it each day and
using it to do a job. Show how it could be improved with a breakdown of
what you could do and what would have to be done by others. What would it
take to make your workplace ideal?"

check list

How many items on the list was the participant able to integrate into one
picture? The participants demonstrate how well they are integrating and
relating material in the course when they are able to exemplify a number of
points in any one picture.

Were the participants able to articulate *why* they decided that a picture they
took exemplifies a particular point on the worksheet?

The photographic presentations could be kept so that if the instructor assigns
this exercise again at the end of the course there can be some measure of each
participant's development.

Nature

15

Working with an ecological model: a field trip is taken to look for order in nature and to begin to define systems as a means of studying nature.

nature is the context of our built environment

The great variety in nature is what makes it endlessly satisfying to watch and to be in. Because of that variety, nature should appear chaotic. It is in fact highly ordered. Each of nature's components — plants, animals, minerals — makes its own patterns within the total nature: a unique, predictable pattern of branches, bark, and leaves enables us to distinguish an oak from an elm. Nature also provides order in the recurring processes of birth, life, and death; in the rhythms of time, day and night; and in the rhythms of season.

The endless variety in nature is also what makes it difficult to study. To do so we must simplify and reduce it from the complex whole, define a group of its parts (making an ecological model or system), and describe limits within which it can be systematically observed. In isolating a system we neglect other parts of the total complex of phenomena called nature. The ecologist has the professional responsibility to be sure the factors he has omitted by setting limits are not controlling variables: to study ecological balance of a lake, he cannot ignore effluents from a nearby industrial plant.

We construct our built environment partly on the basis of what we perceive in nature. Our structures are a part of nature and may be satisfying either because they are integrated with the natural surroundings (a sod house would be an extreme example) or because they contrast with the environment (like a grain elevator on the prairie). In either case, nature is the inspiration for and the context of designed objects and buildings.

In designing landscape, the Japanese construct a garden as Westerners build a house: every plant is mercilessly controlled according to the designer's idea and may be installed at its full size.

The Japanese distill the order they perceive in nature to evoke the spirit of nature in their gardens. Compare this garden with the woods on page 141. What patterns of order do you see in each?

exploration and observation

Let the participants walk in a site free of man-made intrusions for twenty minutes, with the idea of sketching the land and its contents when they return. Work in pairs. When the participants reconvene, ask them to compile a base map containing all the collected information. The group can draw their own base map or use a USGS map.

Point out to the group data not readily observable to add to the map. Topics might include soil types and characteristics, major geological formations with a brief history, types and habits of flora and fauna, amount of rainfall, wind intensities, water systems, and seasonal patterns.

The group should compare their raw data with the contour information on the USGS map. "There may be discrepancies between your observations and the map. Why? How do such changes affect man the builder?"

Ask the group to identify the natural systems that have caused changes in the land-systems such as water, air, vegetation, animals, insects. "The processes of interaction among these systems are also highly significant in causing changes in the land."

Many of the processes can be deduced from the collected data. Although it is not always possible to have an expert present to validate inferences and deductions, what is important is that the participants understand the significance of this body of knowledge and the processes involved in studying nature. It is also crucial that they learn to ask relevant questions.

interdependency of systems

Now ask the participants to go to a new area and inspect it closely. "Enumerate the systems, such as water and soil, then analyze the dependent relationships between the systems, such as rain intensity and soil stability."

Reconvene the group and discuss the interdependency of the systems they observed. Enlarge the discussion to reflect on the interdependency of systems on a much larger scale (regions, continents, earth, and universe).

"An animal's interaction with the environment is direct; man's is indirect, modified by intellect and culture. Do you consider yourself part of nature? What is the difference between our relationship to nature and that of primitive peoples?"

our experience

General information about the area in which you live is available at most city or university libraries. There will be discrepancies between the description of the area in general and what you find on the particular site. Ask the participants to notice the variances.

It is difficult to hold the group together since they are not used to an outdoor situation. They tend to wander away mentally and physically. To keep thinking and conversation focused, discussions are best held in a defined area so that there is some feeling of group closure. Limiting the number of participants in each group to twelve also helps prevent wandering minds.

This field trip must be held in good weather, but not too good. The effects of sun, wind, rain, or a beautiful day overwhelm all other variables.

What is a man-made environment? a bent twig? a torn leaf? a crushed shell? What is our impact on nature? Do we leave a permanent mark or will time always wash away our footprint?

using the concept of systems to help solve problems

Select a location with some variety of terrain, vegetation, rock, and water. State as a problem for the group: "How would one go about 'studying' an insect that you can see on the site and that lives in this 'world'? What limits will you have to set on your investigation?"

The point is to get the group to think in terms of systems as a means of defining and solving problems. One way to do this is to ask: "What are the forces operating on the insect?" (The forces might include weather, available soils, and predators.)

"Where is the insect's shelter? What is his food supply? What is his contribution to the environment? What we are laying out is a description of the insect in his environment, that is, in his total system. We still must decide what aspect of the insect we want to study or research.

"If, for example, we choose to study the physical boundaries of the insect's territory, we will have a complex subsystem of the insect's total environment that includes almost all the factors we have already mentioned. For instance, does the insect run out of his normal territory when man approaches? We cannot discount man as part of this subsystem until we can be sure he has no effect on it.

"If, however, we choose the effect of temperature on the insect's water intake, we can remove the insect to a laboratory cage in which we can standardize the moisture level, food, and light. Holding all the variables constant except the one we want to study, we can change the temperature incrementally and measure the amount of water ingested."

The writing spider stores its food in a cocoon and patches its web in a "writing" pattern. What are the boundaries of its world?

"Let's try the harder problem of studying the insect's shelter. Are there any variables you think we can *ignore?*" As examples are mentioned, ask the group to challenge them. For instance, if one person mentions temperature, another could ask if we can assume the insect doesn't dig deeper if it gets colder. If someone suggests food supply, can we assume that the distance between the shelter and the food source is irrelevant?

"You have now set up such a complicated system that to do research you are forced to make assumptions. For example, we choose to assume (1) there is no influence by man (disturbing the soil), (2) the temperature ranges between forty and ninety degrees Fahrenheit, (3) there is a specified range of rainfall. In other words, we confine our research to shelter under 'normal' conditions. This means we will probably obtain very useful information about the insect by rigorously having defined a subsystem (shelter under normal conditions) of the total system in which the insect exists. It would not be useful information, however, if we wanted to know the impact on insect shelter of a dam to be constructed downstream.

"In all cases of research this process must be gone through. The trick is to be certain a variable that overwhelms all the others has not been overlooked or been assumed constant or negligible. In addition, in later applying the data, we must keep in mind that in obtaining *any* data, we have selected from a whole, and those data probably are not applicable outside the limits we have defined. It is clear why one of the great difficulties in environmental impact studies is predicting, with limited time and funds, the complicated interactions of natural systems.

Yellowstone Park, "nature's wonderland."

choose a variable, play a role

"Nature is one of the most difficult subjects to study. Usually a change in one variable changes most of the other parts. To illustrate this point, each choose a variable to act out; for example, wind, rain, sun, soil, grass . . . ." As instructor, now ask "rain" to choose to fall (indicate the amount and intensity) or not to fall. Ask each of the other variables to predict the effect this will have on them. For example, "soil" will be washed into the "stream" unless "vegetation" holds it tightly. Repeat with the "sun," "wind" . . . controlling the system.

environmental impact

Instigate a discussion of environmental impact studies. "How would you route a highway through this site? How big would you make it? What systems would you select for a detailed environmental impact study? What systems have you left out, and why?"

overnight campout

An overnight campout gives the group an opportunity to see an environment under extremes of light level, animal life, temperature, and so forth. It can also provide a study of man's impact on the "wilderness." After the participants have made camp and settled down, the following topics would be appropriate for discussion. "How could we have better integrated our belongings with the area? How could a designer and builder best translate this 'natural' area into a permanent camping area? What should go undisturbed in that translation? disturbed in a limited fashion only?"

nature: a basic for design

An ideal setting for the following exercise would be the campsite in daylight. "Study an area defined by natural materials in terms of the design elements (color, texture, scale, line, pattern, form, structure, and balance) that are perceivable. Use a camera to record those elements. How do the elements depend on each other for the total affect on the perceiver? Prepare some designs incorporating or suggested by the elements recorded."

nature: use without abuse

Hold the following discussion outdoors. "What impact have people had on this site? Give your answer in minute detail. Wood gathered, leaves pulled from trees, and nuts eaten can be part of the inventory, as well as larger-scale human activities such as dam building.

"How would it feel to discover this area? Is that feeling and the vision that accompanies it an experience valued by many people? Should we preserve natural areas in the form of national and state parks? If so, what controls should be imposed on people's activities in popular parks like Yellowstone and Yosemite to preserve them for ourselves and our grandchildren?" (Factors to consider are vehicular traffic, sightseeing pedestrians and handicapped persons, shelter, food service, toilet and trash facilities, and the special needs of recreational vehicles.)

"Should some places be designated wilderness areas? If so, what level of human activity should take place there? Should overnight camping be permitted? Should the density of plant or animal life be controlled by man?"

Choose an outdoor environment with natural elements such as water, vegetation, birds, insects, and animals. An area with a minimum of man-made alteration is preferable. If possible, the area should have a variety of topographic relief features, for example, hills, plains, dells, or flood plains. Identify on the site either a physical shelter or a spot with natural closure, such as a grove of trees, in which discussions can take place.

a U.S. Geological Survey map of the area to be visited: scale, 1:24,000 with contours

white butcher paper and felt-tipped pens

If possible, include a biologist, geographer, or geologist as a resource person. A scientist has expertise and a point of view to contribute that enhances the exercises.

further exploration

"Using design elements you have found in nature, create something of your own. Put it into a natural context in such a way that it becomes a transition from the natural to the built environment.

"Choose a two-foot-square outdoor area of land near your home. Keep a log, recording two or three times a week all the changes that occur. Continue your log through at least one change of season.

"Read the article by Deevey and outline a series of steps that would enable you to decide how to control the overpopulation of lemmings."

check list

How many "systems" and "subsystems" on this site can the participant perceive?

Can the participant give an example of a "system" other than the ones described, indicating those variables he has included and excluded?

resources

Bedichek, Roy: *Adventures With a Texas Naturalist*, Doubleday, New York (1950). Fascinating observations of wildlife. See particularly Pages 31-39 describing changes caused by artificial reservoirs.

Carson, Rachel: *Silent Spring*, Houghton Mifflin, New York (1962). The effects of our attempts to control our environment.

Deevey, Edward S.: "The Hare and Haruspex: A Cautionary Tale," *American Scientist* (1960) Vol. 48, 415-430. The hazards of trying to control nature.

De Santo, R. S.: *Concepts of Applied Ecology*, Springer-Verlag, New York (1978). A fascinating introduction to the fields of applied ecology and environmental management.

McHarg, Ian L.: *Design With Nature*, American Museum of Natural History, New York (1960). The interaction of man with nature.

"To a Mouse," *Poems and Songs of Robert Burns*, James Burke (editor), Collins, London (1955) 111. Burns wrote this poem in 1785 after ploughing up a mouse in her nest.

U.S. Geological Survey Maps can be purchased from (1) the Department of the Interior, Geological Survey Division, in cities with federal offices, or (2) the Distribution Section, USGS, Federal Center, Denver, Colo. 80225. If possible, describe by latitude and longitude the area for which you wish a map. Otherwise, request an index before ordering, or describe the county, section of county, and any nearby town. Maps of counties and of large cities are usually divided into several quadrants. You may need more than one quadrant to fully cover your designated area.

Sequence of spaces

16

Building an environment for the study of contrasting spaces: the group, working in teams, plan and build a series of contrasting spaces out of cardboard.

all sensory perception is achieved by contrast

We perceive and remember the shape of one space by its contrast with another. Built spaces are viewed as "large" or "small" in relation to spaces leading to and from them. Surprises in experiencing spaces come from deliberately strong contrasts between the spaces. Our emotions are affected, consciously or unconsciously, by the varying qualities of space around us. The shape of space is one of the environmental qualities that affect our emotional response. It is perhaps the most important visual quality to consider in designing a building or group of buildings.

The *process* of decision-making is particularly cumbersome in designing physical environments. Decisions generally are made by groups and often are affected by conflicting personalities, pressures, and points of view. These human processes affect any physical product or project. They are a continuing part of the product. Human interaction with the product continues as the product is built and used in an ongoing, evolutionary process.

Fatehpur Sikri was a Mogul new town in 16th century India, abandoned for lack of water. Red sandstone pavilions surround open courtyards forming a sequence of spaces glimpsed through arches and stone grills.

a walk through contrasting spaces

Lead the group on a walk through a sequence of connected, contrasting spaces. Encourage them to explore and to "feel" the different impact of each space. "Remember, you are always 'in' a space whether you are inside or outdoors. As you move from one space to another, what effect did the first space have on your reaction to the second space? What factors are present in each space that cause you to alter your pace?"

In the meeting room, ask the group to sketch diagrammatically the sequence of spaces they have just experienced. "Show how these spaces affected you and describe why. Did some spaces relate to each other better than others? Why?"

building a sequence of spaces

Outline the project to the group, emphasizing the following: "This is a team project. Working together, build out of cardboard, and any other materials you wish to supply, a continuous sequence of spaces to illustrate the concepts we have been experiencing. The spaces should be large enough for a person to go through."

Decide and state to the group the basis on which the project will be judged: design, cooperation within the group, participation by individuals. To help focus the process, set specific goals for the class and for yourself.

You can give structure to the activity by outlining the exercise on the board and setting a time limit for completing each stage. For example:

1   Discuss how different effects can be achieved by using contrasting spaces in varying sequences.
2   Form teams of three or four, according to your choice.
3   Decide with your team what your space is to be, how it will be built, and how it will contribute to the whole sequence.
4   Specify in writing the intended effects of each part of the sequence of spaces.
5   Review (as a group) the design of the total project with the help of drawings on the board. (Any modifications made during the building should be recorded on the original specifications and kept for reference after the project is completed.)
6   Build the sequence of spaces.

Emphasize to the group that they must consider the relationship of one space to those adjoining it and must consider the whole from the point of view of pacing one's way through it.

Point out that the overall labyrinth must be inviting. Too much crawling on the belly, moving in the dark, or changing levels has an overwhelmingly negative effect on those going through it. Too many extraneous objects — such as pillows and plants — tend to diffuse the impression of the space itself.

our experience

It is helpful to establish a major goal — is it design, group problem solving, or the effect of a group process on design decisions? Time limitations may not permit stressing both "process" and "a design solution." Make your decision clear to the participants.

You will also need to clarify your own role to the group. Will you act as the authority, as the final arbitrator of their decisions, or as the disinterested facilitator?

It takes the group considerable time to make the decisions that get them started on designing and building the spaces. University students, particularly, argue more than high schoolers and adults. Participation in the group dynamics of this project is one of the most valuable experiences mentioned by former participants.

About halfway through building the spaces, many want to alter the execution of their idea, or change the idea altogether. It is up to you whether to allow this or to commit the teams to their original schemes.

experiencing the labyrinth

Ask the participants to remove their shoes and go through the completed sequence of spaces — one at a time. They should go slowly and remember what they experience and how they feel about it. (Visitors should not be given any clues as to what they will find.) As each participant comes out of the labyrinth ask him to jot down his reactions to the experience as a whole.

"Try to recall as many of the spaces as possible. What were your emotional reactions to each space? What did each make you think of? What adjectives would you use to describe each?

"How are you able to remember the spaces? What effect did the sequence of the spaces (high to low, light to dark) have on your ability to remember them?

"Try to draw a plan of the labyrinth from memory. Do you find it hard to draw?" (Everyone does because it is difficult to communicate three-dimensional information on two-dimensional paper.)

analyzing the experience

As each participant completes his sketch, hand him a prepared diagram of the labyrinth. As groups of about eight participants finish, suggest they find an appropriate place near the labyrinth to hold a discussion. Include in each group one or two visiting instructors who have experienced the sequence of spaces.

"Compare your sketch of the sequence of spaces with the accurate drawing. Try to analyze whether the differences between your sketch and the diagram were caused by such qualities as lighting, odor, noise levels, and color, or your own 'feeling reactions.' " (anxiety, claustrophobia . . .) "Perhaps some of the spaces and the way they were juxtaposed were not very memorable. Why not? Are there any spaces that none of you remembered? that all of you remembered?

"What senses did you use in perceiving the spaces? Did each space usually appeal preponderantly to one sense, such as touch, smell, or hearing?

"In which space would you choose to spend the greatest amount of time? be alone? have a conversation? sleep? Why? In which space did you feel the most comfortable with respect to size?

"What variations in movement, pace, and levels do you recall? Did the design of the labyrinth control the rate of your movement through it? What changes in level could have been used to get more variety of kinesthetic and proprioceptive experience and thus enhance the total experience?

"Imagine the labyrinth had been shaped out of living plants instead of cardboard. What do you think your reaction would be then? Would your reaction to your least favorite space change? Why? If you had to do it frequently, would you be more bored going through a cardboard labyrinth, or through a plant labyrinth?"

the shape of space

"Although we have purposely exaggerated the shape of space in this exercise, everyone experiences a variety of spaces daily, though often unconsciously. In a dramatized experience such as building a personal space or a sequence of spaces, you may more easily recognize that space does indeed have shape. When you become aware of the shape of space you will be able to judge its psychological effect on yourself, as well as its functional efficiency.

"Tour the outside of the sequence of spaces, matching the exterior forms with the interior spaces. Are the exterior forms as you imagined them while going through the labyrinth? Are there any surprises?"

Show the group a solid model of the sequence of spaces (or of a simpler space) such that the walls may be removed and a solid left that fills the internal volume created by the walls. Discuss the mental tool of visualizing the shape of space apart from its container. "A good way to mentally visualize the shape of space is to imagine it poured full of plaster of Paris and then the walls removed. Outdoor space also has shape. A street is shaped by rows of building facades or trees.

"Name some other examples of exterior or outdoor spaces. How are they defined or contained?" (yards, parking lots, parks, and courtyards, defined by building walls, plants . . .) "What contains these open spaces? What forms are their boundaries? Do the same qualities apply to outdoor spaces as to inside spaces?"

Ask the participants to go outdoors and sketch on paper the shape of spaces formed by a series of buildings you have previously identified. When the group returns, discuss the difficulties of communicating voids. Reemphasize the powerful effect the shape of space can have on one, reminding them of their own reactions in the labyrinth.

the shape of space looking at a plan:

The walls of buildings define linked outdoor spaces of contrasting sizes and shapes. The periphery of this impulse shopping center is defined by planting and low walls on one side, and by arcaded store fronts on the other. Note the relative sizes of the spaces meant for cars and those planned for pedestrians. Several of these spaces are illustrated on page 144.

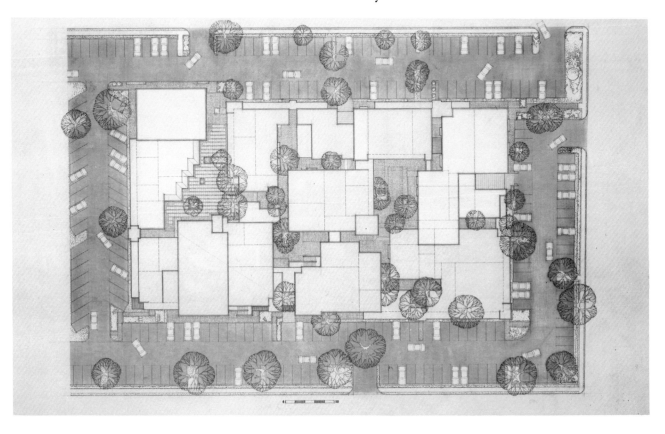

analyzing the process

The frustrations of decision-making by a group surface frequently during this exercise. After the sequence of spaces has been built and the final product experienced, a discussion of what effect the group dynamics and problems had on the final decisions and product will help clarify the mechanisms of working in a team.

"How were the decisions made in this project? by vote? by force of one or two personalities? by asking the instructor to decide?

"Did you consult your neighbors before making the final design decisions for your space? Did you review your progress to see how your section of the labyrinth fit with the whole? Did you have to modify your space in any way?

"How as a team did you deal with other teams? What was your effect on your neighbors' space and vice versa? How did you handle a conflict with your neighbors? with the whole group working on the project?

"In what ways did the group process and problems dealing specifically with people affect the physical product? Would the project have been different if the group interactions had differed? Compare the realized product with the written intentions.

"Why haven't the non-participators 'done their share' of work? Did the teams participate equally in the decision-making? Did the members of a team participate equally in the decision-making? If not, why? What factors precipitated a final decision? Was this decision carried out, or was it substantially modified along the way? If it was modified, why? Did someone 'take over'? If so, why did the rest of you let this happen?"

the shape of space looking at a model:
The center void is an exterior space with a defined "gate" formed by eight attached residences. Glass walls define a private outdoor space within each house visible in the model as a hole in the roof. Trees and plants form small spaces surrounding the complex. A plan showing these dwellings in their neighborhood context is shown on page 119.

Identify a sequence of interior and exterior spaces having as much variety as possible.

Materials for building the Sequence of Spaces, based on twenty participants:

double-faced corrugated cardboard — about thirty 5 feet by 7 feet sheets

duct tape — six 100 yard rolls

matte knives

measuring devices and straight edges

white butcher paper

translucent fabric or bed sheet — about 250 sq. feet

Optional materials to be supplied by participants: (The instructor should limit the number and variety of materials.)

tactile materials such as carpet, tile samples, bricks, fabrics

olfactory materials such as hay, natural vegetation, artificial scents

sound equipment including tapes of common sounds at both usual and unusual volumes such as traffic, water, trains, air hammers

lights, and colored tissue paper

Guest instructors can be used throughout this project to help focus its planning and execution. Their different points of view concerning the significance of the process and the product adds a valuable dimension. They can also function as professional critics to analyze the completed labyrinth.

50 participants or more can work together on one Sequence of Spaces. Classes of several sections can be combined for this project.

further exploration

This experience can be extended according to the interests and needs of the participants:

Designate an outdoor space such as a nearby public park. Ask the participants to analyze and be able to report their feelings as they experience this space. Ask them to list all the qualities they can about the space. Emphasize that different people perceive in different ways and that there are no "right" answers.

"Look carefully at the shapes of at least five different spaces inside and outside your school environment. Do the spaces function well for their purpose? Are their boundaries well defined? Which senses do they best satisfy? Which do they satisfy least? What qualifying adjectives can you attribute to each of the spaces?

"Look carefully at your home environment. Which spaces function best for their purpose? Which make you feel good? Do the outside spaces relate well to the inside spaces? Which is your favorite space? Why? Which do you like least? Why?

"Prepare a photographic essay of the labyrinth."

Architecture students can have an invited jury critique the sequence of spaces. They can graphically record the sequence of spaces.

Psychology students can design adaptation experiments to be carried out within the labyrinth.

check list

Was the participant consciously aware of and able to suggest manipulations of the boundary shapes that define both interior and exterior spaces?

Was the participant consciously aware of and able to express how light, color, and texture influence the shape of space?

Can the participants articulate the effect of the group process on their design solution?

resources

Bacon, Edmund N.: *Design of Cities*, Penguin Books, New York (1976). See Pages 95-111 for a lucid analysis of sequences of spaces.

"Ii centro laico di Fatehpur Sikri," *Architectura* (April 1976) 746-751. Illustrations and plans of this intriguing architectural complex.

Leonard, Michael: "Humanizing Space," *Progressive Architecture* (April 1969) 128-133. Excellent discussion of movement through a sequence of spaces.

Once the target audience and media are chosen, the participants should be required to stay with their choices. The projects are an integral part of the course. They are not term projects done outside the class, although outside time will doubtless be necessary. Part of the teaching strategy in many sessions is collection, analysis, and translation of data and experiences for the project. (For example, those participants doing slide shows or films will be photographing during field trips.)

The instructor needs to monitor the projects closely. As the experience with the Sequence of Spaces indicated, participants will probably "waste" a lot of time forming teams and crystallizing ideas. That in itself is a valuable learning experience. The participants should be encouraged to analyze the group dynamics, perhaps using videotape. It is up to the instructor to prevent this process from going on too long and from becoming a defense against decision-making and action.

living in community: an optional long-term project

Chapters 1 through 16 have dealt primarily with concepts and experiences scaled to the individual in his day-to-day environment. The Sequence of Spaces (the labyrinth) has been used as a principal means for the participant to integrate and apply the concepts and experiences he has begun to explore.

The latter part of the book, Chapters 17 through 27, deals with the more complex problems of the larger urban environment. These chapters incorporate an optional long-term project designed so that each person can work individually or as a part of a small team.

The participant is expected to apply the principles dealt with in the first part of the book to the larger-scale urban problems dealt with in the second part. The project and techniques are designed to cause him to constantly synthesize his experiences into wholes while simultaneously analyzing the parts. The goal is to teach a disciplined ordering and relating of experiences, using the environment as a learning laboratory.

The participant should design his long-term project to communicate the principal themes and insights of the book to an imaginary, uninitiated audience. Having selected his own target audience and communication media, each participant should be required to include the significant concepts of *each* chapter and to explain them. Examples of media might be a text or lesson plans for children, a slide show for adult education, a presentation for a city council, a multi-media show for college students, or an interpretation of environmental problems to be used as a museum exhibit.

17

Personal space and territory

Psychological possession of physical space: participants measure their space bubbles in exercises that demonstrate how our use of space relates to our psychological and social needs.

"some thirty inches from my nose

The frontier of my Person goes
And all the untilled air between
Is private *pagus* or demesne.
Stranger, unless with bedroom eyes
I beckon you to fraternize,
Beware of rudely crossing it:
I have no gun, but I can spit."
W. H. Auden
Prologue: *The Birth of Architecture*

The boy signals with his head that the girls are too close; they are inside his space bubble. By leaning back, the girls acknowledge that they also feel too close. The averted head, and forced smile express the discomfort the shopper feels. She resents this invasion of her space bubble, but is too polite to say so.

As individuals, we require an area of personal space around us wherever we are. This personal space may be visualized as a bubble surrounding us as we move. Its size depends on our personality, age, and culture. Our space bubble expands or contracts according to our needs at that time and the social context. We step away from an enemy; we embrace a close friend.

Personal space, according to Edward T. Hall, may be viewed as serving four needs: intimate space (0 through 18 inches), personal space (18 inches through 4 feet), social distances (4 through 12 feet), and public distances (12 through 25 feet or more). Unsolicited invasion of our space bubble affects our feelings and, therefore, actions, though we may not always be consciously aware of what is bothering us.

The size and configuration of our space bubble affects how we perceive and experience a space. Persons with large space bubbles arrange their offices with their desks as barriers to intimacy. They feel uncomfortable when forced into close personal contact by tightly arranged seating areas. These persons perceive their offices as "roomy" and the tight seating areas as "crowded"; the two spaces may actually be identical in size.

The appropriate design of containers for social interactions requires awareness of the personal space requirements of the users. If facilities do not provide enough space to support our space bubbles in a particular social context, we become stressed. We "can't sit still." These considerations should be part of the criteria by which we judge how well the design of a built environment functions.

In addition to having a space bubble — an extension of ourselves that we carry with us — we seem also to have territories. However, our "possessed" territory is not carried with us; we occupy it, instead, when we move through it. We develop a number of territories related to our habitual routes to work or school. Some of those territories, such as the sidewalk in front of our home or our parking place at work, are shared constantly. Others may be shared at our will, as, for instance when we invite a neighbor into our yard or home.

Territory can be captured even momentarily. *At upper left* a woman has claimed a park bench with packages and extended arms. The man avoids recognizing this claim by backing onto the seat. *At upper right* each person resting on this stair maintains his space bubble by avoiding eye contact with anyone nearby. *At lower left* an island territory for two is staked out with newspapers and purse. Would you dare to take the center chair? *At lower right* extended limbs and a blanket establish territorial limits.

Our sense of well-being is enhanced when we perceive clearly our territories and our relationship to them. The intensity of our personal responsibility for what happens to those shared territories increases with their closeness to our home and the amount of time we spend in them. We protest vigorously the removal of a tree from the parkway in front of our house. We observe without comment the loss of a unique stand of trees across town. We have a fragmented view of a city or an area because we tend to observe and care about only that part of the city we perceive as "our" territory. Twentieth century mobility further exacerbates our fragmented view of city. We move so frequently we do not spend time enough in one place to identify with it strongly. All of these factors hinder our ability to view our city and ourselves as an interrelated whole.

discovering space bubbles and staking out territories

Ask groups of about twelve participants to tape butcher paper to the floor, forming a square approximately ten feet by ten feet. Ask the participants to remove their shoes, and all simultaneously walk around exploring various places to stand on the paper. Ask them to sit down when they have found the place on the paper where they feel most comfortable. Give each participant a black felt-tipped pen and ask him to consider, then mark, the boundary of the intimate space he needs for himself.

Devise an exercise to relax the participants and focus them on themselves and on their own space. For example:

"Close your eyes and imagine that you have a third eye with which to explore your own body. What does it look like inside your knee, your ear, your throat, your spine? Place your new eye in your stomach and remember the first two words that come to your mind. Open your eyes and draw in your space what you saw. Add the two words."

Ask the participants to "Stand up and find the place and stance in your space that pleases you most. Outline your feet. With whatever colors you like, personalize your space by drawing more in it.

"Trade spaces by selecting another space that is comfortable to you. How do you feel about this new space? Why? How do you feel about that guy standing in your original space? Why? Trade spaces again. Select the one you feel most comfortable in. Is this one just the right size? What changes would you make? What would you add or subtract? Outline your feet and make the changes. How do you feel about this space now that you have changed it?

"Look at the person standing in your original space. What is he doing to your space? How do you feel about his changes? Why? Move off the paper, walk around it, make as many observations as you can about what is happening on that paper."

Discuss the participants' observations, always asking them for the reasons for their statements. "Do some of you prefer a corner space or one near a door? Can you analyze why? Do you also pick a protected seat or an escape seat in a movie theater or meeting room? Did you stake out your territory next to your friends' spaces? If so, when all of you traded spaces on the paper did your whole community of friends move as a group? Do different people's spaces vary in size? Measure some and record their dimensions. Were you comfortable in the amount of space you had or did you want more?"

our experience

This exercise illustrates how people identify with "territories" and relate to each other spatially.

This is a good occasion to discuss the use of space and the human need for it. Compare how space is used in parks, offices, schools, freeways, public rallies, jails, and airports. The experience can be a springboard for all sorts of psychological and sociological discussions if you choose to use it that way.

If you, as the instructor, participate directly in "discovering space bubbles and staking out territories" the territories will all tend to be related to you as the authority figure.

finding the size and shape of space bubbles

Unobtrusively begin to invade someone's space bubble. When the participant moves to "escape," point out what you have been doing. Suggest they try this exercise with strangers sometime and note the reactions they observe.

Ask a volunteer to stand in a marked place while another volunteer approaches him from the front. Ask the first participant to indicate the distance at which the second has come "too close for comfort," that is, at which he begins to feel uneasy. Mark that distance with chalk. Ask a third volunteer to approach the first from the side until the first feels uncomfortable and again mark the distance. Repeat, approaching from the rear, and sketch roughly the contours of the space bubble thus defined. Ask one or more other participants in turn to stand in the center, mark, and repeat the exercise.

After obtaining space bubbles in several sizes, begin a discussion: "If people's spaces differ in size, what are some of the possible reasons why this is so? Does a person seem to put greater distance between himself and another person standing in front of him, behind him, or on each side of him? What are some possible reasons for this? Develop some hypotheses for this phenomenon and test them against your experience."

invading intimate space

Lay out butcher paper on the floor to form a rectangle simulating the shape of a city bus. The area should be large enough for the whole group to stand on, but so small that they are crowded together into physical contact.

Ask the group to board the "bus" for a ride, saying: "You are on a very crowded bus, one full of strangers. It is rush hour on a Friday evening. This ride with these strangers will take fifteen to twenty minutes. Think about and try to remember — in general, how you feel. What would you like to do at this moment?"

Ask the group to be perfectly still. At some point (which may occur so quickly that you must watch closely to detect it) someone will say or do something. Note the length of time it takes to break the silence. Ask the participants to get off the bus. Take a short break, then reload the bus. This time board only as many people as feel comfortable for a fifteen-to-twenty-minute ride. Remind them that it is still rush hour and that they are to stand perfectly still. Again note the amount of time it takes to break the silence.

Discuss the differences between the feelings on the first bus ride and those on the second. Compare the time it took for the silence to be broken on the crowded trip with that on the uncrowded ride. (Anyone's anxiety is greater when his intimate space is being invaded. Individuals will deal with this increased anxiety in many ways, depending on the situation, and on their personality, culture, and age. Two fairly common ways of dealing with this increased anxiety are to reduce it with some friendly communication — verbal or body language — or to reduce it with some angry verbal or nonverbal gesture.) "What are the implications of human needs for, and use of space in other crowded situations, such as stadiums, jails, and airports?"

A passenger on the "bus" clowns a pratfall to break the tension caused by overcrowding. Notice that everyone has his hands contained or concealed.

identifying territory in the city

Ask each person to draw a freehand map on a piece of butcher paper, marking his residence. "Locate the places you frequent daily, drawing the movement corridors (walks, streets, trails) you use to reach them. Next add places you use periodically — every week or month — that are familiar enough to evoke a feeling 'this is my territory, too.' Add the less-often-used movement corridors in a different color. Finally, add places used infrequently — perhaps visited several times a year — but familiar enough that you feel some connection with them. Add those corridors in a third color."

When the group have finished their freehand maps, ask them to imagine a "nice place" along a corridor they frequently travel. "Suppose you drive by there on your way home and see the large sign of a developer 'For Sale, Will Build to Suit.' Another favorite place has been torn down for urban renewal. Do you feel at least a twinge of personal affront or loss? Why? You know logically that the place is not yours, so you have no say about it. Then why the twinge of concern? Or you get to school a little late and your usual parking place has been taken. You know it is irrational, but there's a slight irritation: 'Who's got my place?' "

Ask each participant to transfer his freehand data to a standard city map. "What are some of the obvious differences in the configurations on the two maps? Do you find that streets you use frequently are shortened in your mind, and therefore on your freehand maps? Compare each other's city maps. Are there territories that some of you have in common? When one thinks of his city, does he think of it as a whole or as the fragments that compose his territories? For the rest of the course we will be confronting urban problems that develop from the fragmented way people relate to their city. In a sense, this fragmented way of perceiving is 'object oriented' as opposed to 'field oriented.' "

our experience
People are often completely unaware of their territories. You may want to return to this exercise in several days, after the participants have had time to recognize their territories as they travel about in their communities.

Graffitti once marked this wall as Thomas Jefferson High School territory. A mural painted by the students now makes the same territorial statement but also enhances the street.

preparation

black and colored broad-line felt-tipped pens

white butcher paper

1" masking tape

chalk to mark on floor

measuring device

city map for each participant

watch with second hand

further exploration

"Experiment with the concept of space bubble by quietly invading another person's bubble at a party or while waiting in line.

"Look for evidence of territorial space in a library and in an eating place you frequent."

check list

Was the participant able to recognize when his space bubble was invaded by another person?

Was the participant able to discern and articulate his own use of territorial space?

Could the participant identify and articulate appropriate use of the concepts of personal space and territory in environmental design?

resources

Birdwhistle, R. L.: *Introduction to Kinesics*, Foreign Service Institute, Washington, D.C. (1952). Systematic studies of how individuals perceive, and orient to, one another.

Environmental Psychology: Man and His Physical Setting, H. Proshansky, W. Ittelson, and L. Rivlin (editors), Holt, Rinehart and Winston, New York (1970). See particularly the article by Stea (Pages 37-42) on space, territory, and human movement.

Hall, Edward T.: *The Hidden Dimension*, Doubleday, New York (1966). An anthropologist discusses man's sensuous perceptions of space and how he uses space in public and private places.

Hall, Edward T.: *The Silent Language*, Doubleday, New York (1959). The use of space and time for non-verbal communication by people of different cultures.

Morris, Desmond: *Manwatching: A Field Guide to Human Behavior*, Harry N. Abrams, New York (1977). An abundantly illustrated journal of human behavior.

Sommer, Robert: *Personal Space*, Prentice-Hall, Englewood Cliffs, N.J. (1969). The effects of physical setting upon attitudes and behavior.

Personal needs

18

Multiple choices in lifestyles: discussion and field trip reveal the physical, social, emotional, and other needs that must be met in planning for individuals living in communities.

total independence is an illusion

We each have basic physical and psychological needs. Each of us chooses how we will meet these needs. Our pattern of choices becomes our lifestyle. Although we can satisfy some of the needs by ourselves, we are dependent on other people. Our choices are to an extent defined by the community of people in which we live and the alternatives available, such as existing institutions and the activities associated with them, and the community's tolerance for deviation.

The needs of individuals of varying age, culture, income, and temperament are different. Therefore multiple choices of lifestyles are desirable. Each of us benefits vicariously from having in our milieu this rich diversity of living patterns that can touch our lives and enrich our experience.

The physical environment is the container for our experiences. Man has proved remarkably adept at living in whatever environment he finds himself — palaces, prison camps, permafrost. The enormous elasticity and adaptability characteristic of man has served him well in extremely adverse circumstances. This ability, coupled with the social mores that imply "man's lot is to suffer," at times works against him. We tend to think we don't merit a choice — that we can't reasonably expect to have all our physical, social, and psychological needs met. Such feelings can prevent us from seeking alternative solutions that would more fully satisfy our needs and enrich our lives.

Three different lifestyles have in common a relatively small living space in a community of similar dwellings. *In the center* pre-industrial Isphahan, in 1970 still expressed a lifestyle dependent upon animals and feet for movement. Compact dwellings designed for protection from the desert climate leave only courtyards open to the sky. Even the streets (a line of domes) are covered for protection.

On the left, a mobile home park in 1977 makes a transient lifestyle possible, although in reality the homes are infrequently moved.

On the right an American suburb in 1976 illustrates the low density of single family houses made possible by private cars and Federal Housing Authority mortgages.

The same meal takes on a new character in each different setting. The context makes the difference.

comparing lifestyles

Ask the participants to spread out the maps of their territories prepared during the previous session. Start a discussion to compare the various lifestyles of the participants.

"How much time do you spend in activities at specific locations on your map? Where do you eat each meal? always at the same place? Where do you shop? What do you do for entertainment? for recreation? Where do you go for those activities? Do they take place at home? with family? with friends? Could you do these same things alone? Do you think a hermit could be *totally* independent?

"How much time do you spend in a car?" Ask that each participant keep a log of the time he spends in his car each day for a week. He should also record the time he spends in *public* spaces. The data will be used in the sessions on Movement Systems and Public Spaces.

"Recall where you lived when you were ten years old. What kinds of things did you do outside school hours? Where did those activities take place? What things couldn't you do in your territory? How did your family's activities differ from those of your close friends? Imagine yourself as a ten-year-old where you live now. Would you have more or less freedom than you actually had as a child? What makes the difference? What is missing or added for the child?

"Those of you who know grandparents or great-grandparents, pinpoint their neighborhoods on your map. In what ways do older people live differently from the way you do? Do they drive cars? If not, what limits do they have on their mobility? Do they own a home? Do they leave home for their entertainment? Imagine them in *your* present living conditions. What would they lack, gain, or have and not need?

"Look at your own neighborhood. Can you say that the people who live there have a lifestyle in common? If so, what are its characteristics? If not, describe the extremes of lifestyles you have observed."

"We are not usually aware of the great differences in the ways even our own friends and associates live. Let us look at lifestyles that we have around us or that we know." Select examples appropriate for the particular participant groups. For example:

"*The student living on campus in a dormitory:* How well is his need for privacy met? for sound control? for personalization of his immediate environment?

"*The student living at home:* To what extent does the physical environment of the home make this student's life different from that of the dormitory student? What changes does a resident college student cause in his home environment? In what respects does the academic culture of the university conflict with the family culture?

"*The married student:* Does the married student live in a student community or is he isolated from his peers? Does his lifestyle differ from that of a single student? What special requirements, such as day care for the children, does the married student have?

"*The family:* What visible differences are there between the lifestyles of low-, middle-, and high-income families? Are the differences greater between different races or between different economic groups?

"*Ethnic cultures:* Are there customs of different cultures carried on in your community that are supported or limited by the environment? (The Latin *paseo*, that Sunday evening promenade around a park, is often not provided for in communities built to satisfy Anglo needs.)

"What are the two most distinctly different family lifestyles represented in this group? How could you enhance the way you live within your family? If your grandparents live with your family, in what ways could their personal needs be better satisfied? Visualize how you will be living ten years from now. In what respects does it differ from the way you live now?"

observing lifestyles

Lead the group on a walking or automobile tour through an area of varied lifestyles. Ask the participants to begin their long-term group projects by illustrating a diversity of living patterns, characterizing each by the physical environment that they see, hear, smell, and touch.

"What are some of the visible characteristics of the ways people live? Look at both the physical environments and the social interactions. Make the notes, sketches, or photographs needed for your projects and return to the meeting room with them.

"What criteria did you find useful to discern differences in lifestyles? Can you use the same criteria to suggest how a particular lifestyle might be enhanced? What influence does the *design* of the physical environment have? You can *adapt* to almost anything: *you* change to fit your environment. What examples did you find of people having changed their environment to satisfy personal needs? What kinds of contact did you find between neighborhoods having populations with different lifestyles?

"Will you give the other person the right to live his lifestyle? What if it conflicts with your's?"

The lifestyle of these Moslem women is reflected in their dress. They have, in a sense, changed the environment of their space bubbles by enclosing themselves completely from public view. How would your personal dwelling change if you adopted this concept of privacy?

As a follow-up for this field trip, ask the participants to help write a questionnaire to use in the next session: "The word 'neighborhood' describes that part of your territory in which you live and buy at least convenience foods." Pass out copies of the "neighborhood questionnaire" and encourage the group to add questions to it. "To better understand what makes a successful neighborhood, we need to observe the ages of the residents, their income levels, and their lifestyles. What questions will help us analyze how each of these factors affects the character of a neighborhood?"

neighborhood questionnaire

1   What factors make you think of this as a neighborhood?

2   How does this neighborhood relate to the city as a whole? How easy or difficult is it for the residents to get to the city as a whole? How easy or difficult is it for the residents to get to the other parts of the city they use? To what extent do city transportation routes go through the neighborhood? How is this handled? What other systems does the neighborhood share with the city? (alleys, creeks, power lines)

3   What did this neighborhood look like ten years ago? twenty years ago? What evidence leads you to your conclusions?

4   What relatively permanent structures and patterns in this neighborhood would influence strongly its shape and use if the homes were razed for new development? What trends can you see that suggest how the neighborhood will look in ten years?

5   What kinds of symbolic and physical barriers to calling, walking, or sharing gardens between dwellings do you notice?

6   What are the common denominators that tie the dwellings together visually while still allowing some freedom of choice to the owner?

7   How well does the neighborhood meet the lifestyle needs of the people who live there? Are any elements lacking? How could those needs be supplied and financed?

8   What are the individual's public responsibilities and private rights as a resident of the neighborhood? Can you point to any examples of conflicts between these?

9   Is there any place in this neighborhood that you feel particularly comfortable? Can you say why or why not? What changes would it make in your life if you moved here?

10   If an archeologist a thousand years hence visited this area, what impressions do you think he might receive? What evidence will remain? What conclusions might he draw about these people from the physical evidence?

11   What noises and odors are you conscious of in this neighborhood? Do any of them enhance the character of the neighborhood?

12   Find evidence of the different city service systems (telephone wires, street furniture, fire stations) and note how they relate to the needs of the residents. Which systems are well handled? Which are not? What service systems do you think this neighborhood lacks?

13   Where do residents of this neighborhood shop? for food? for shoe repair? Where do the children buy candy and ice cream? Is the shopping environment visually integrated into or set apart from the neighborhood? How can the backsides of commercial buildings and the activities that are usually carried on there be arranged so as to have a positive effect?

14   In what respects is this neighborhood different from the neighborhood in which you live?

preparation

maps prepared in Personal Space and Territory

Identify geographical areas having a diversity of lifestyles.

The logs requested as part of "comparing lifestyles" will be used to compare time spent in the car in Chapter 21 and time spent in public spaces in Chapter 24.

a copy of the "neighborhood questionnaire" for each person

further exploration

"Select an area that represents a lifestyle markedly different from your own. Observe this neighborhood over a period of time, going back at different times of day to see how people use it. Is the environment of this neighborhood appropriate for the functions you observe there?"

check list

Was the participant able to describe his own lifestyle in comparison with that of someone of a different age group?

Was the participant able to express changes he could make in his environment to facilitate and to enhance his personal way of living?

Was the participant able to communicate 'lifestyles' in his project?

resources

Cooper, H.S.F.: "Life in a Space Station," *The New Yorker* (Aug. 30, 1976) 34-70; (Sept. 6, 1976) 35-70.

Defoe, Daniel: *Robinson Crusoe.* Description of adaptation to a new lifestyle.

Golding, William: *Lord of the Flies.* Terrifying lifestyle developed by children shipwrecked on a deserted island.

Goodman, Paul and Percival: *Communitas: Means of Livelihood and Ways of Life,* Vintage Books, New York (1960). Witty, readable look at city plans of the past and alternative plans for the future.

Griffin, John Howard: *Black Like Me.* How do you ever "know" what a vastly different lifestyle feels like? An engrossing account of the experiences of a white writer who temporarily became black.

Lewis, Oscar: *Five Families,* Basic Books, New York (1959).

Lewis, Oscar: *La Vida,* Random House, New York (1966). Both of these books show how an anthropologist is able to observe and record lifestyles. *Five Families* immerses the reader in the lives of families living in Mexico. *La Vida* is a fascinating description of a family living in Puerto Rico and New York City.

Netzer, Dick: *Economics and Urban Problems: Diagnosis and Prescriptions,* Basic Books, New York (1970).

19 Neighborhood

Toward understanding the issues of neighborhood design: a field trip is conducted to observe a neighborhood within the context of the city as a whole.

our neighborhood is shared territory

"Neighborhood" describes that part of our shared territory in which we have our dwelling and usually spend a great deal of our time. Some people visualize their neighborhood as a geographical area that includes the shops where they buy convenience goods and services. Its boundaries may be major streets. Others see their neighborhood radiating from their homes to any number of close or distant points. This viewpoint is made possible by the car.

However we define our personal neighborhood, our lives can be greatly enhanced if the streets, walks, parks, and stores are designed to facilitate our daily activities. Our identification with physical "place" determines our sense of community. It also determines the degree of personal commitment to maintaining our neighborhood as a pleasant place in which to live.

Each neighborhood impinges on and is in turn influenced by the fabric of the whole urban area. A great diversity of neighborhoods in a large city is stimulating and exciting. Many problems can also arise from the inhabitants' differing values. One group may favor rapid growth; another resists change.

Before the nineteenth century most dwellings were organized in neighborhoods within walking distance of the village square or main street. Around 1900, Ebenezer Howard and later, Clarence Stein, proposed neighborhoods with defined physical boundaries, school, and self-contained convenience goods as a refuge from industrial areas. Today a neighborhood is, for many people, a network of services linked at any distance from their dwellings by their private cars.

All neighborhoods exist in time: past, present, and future. A new neighborhood can be an asset to a city. How well the neighborhood handles change over the years determines whether it is an asset or a liability to the city it is a part of. The transition from a cotton patch to a planned suburb can represent an overnight erasure of the past. However the developer, rather than bulldoze all traces of the past, can choose to incorporate existing land contours and vegetation. In older neighborhoods, the builder can choose to relate his new projects to the existing structures. In this way the present and the past are physically integrated. The best of the past can be preserved while new development revitalizes the area. The neighborhood as a whole is enriched if the new designs are sensitively incorporated with the old.

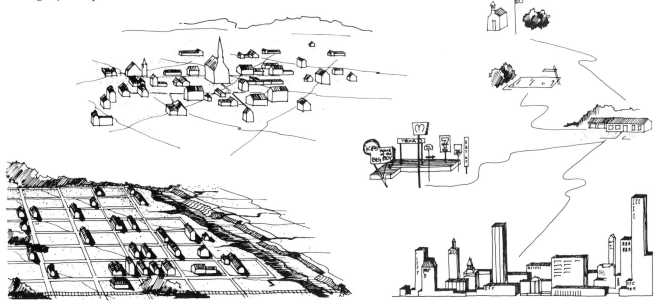

analyzing neighborhoods

Ask the group to meet at a designated place in the neighborhood with their questionnaires. "Work in teams of two or three persons. Write down your observations and answers to questions as you walk and ride through the neighborhood."

The questionnaire may be used as a point of departure to begin the subsequent discussion. "How can a city serve the needs of its people? In previous sessions we have discussed territoriality and other basic human and personal needs that people seek to satisfy in carrying out their lives. Perhaps one of the most frequently overlooked needs is that of people to be a part of the planning and doing process, to have some control over the processes that change their living space.

"You have just observed a neighborhood. Relying on your experiences during the field trip and previous discussions, begin analyzing the neighborhood. Use butcher paper freely to take notes and sketch. Describe how this neighborhood functions. What does it contribute to the city as a whole? Does the design facilitate or inhibit its functioning? Is there a central focus or meeting place such as a shopping center which defines the neighborhood? Use some old pictures. and maps of the neighborhood you visited in order to understand something about the process of change in a neighborhood. People have lived in many different ways in this neighborhood. Can you characterize and say what determines the various living patterns? Use your imagination to visualize the people using the neighborhood. As this neighborhood changes, will it continue to make the same contributions to the city as a whole?"

Arrive at a working definiton of a neighborhood to bring the discussion to a close and to lead into the exercise "proposing a neighborhood." "Each of you visualize your own dwelling. Now think of yourself walking or driving from your dwelling into your neighborhood. Continue your imaginary walk or drive until you have covered all of your neighborhood. What pattern does it form? Does it have fixed boundaries? Is there a central locus that makes it *your* neighborhood? Has your mental pattern of the points or limits of your neighborhood changed over time, even if you have stayed in the same dwelling? Is there a difference between 'neighboring' and 'neighborhood'? How shall we define 'neighborhood'?"

our experience

If time permits, select several neighborhoods to visit. Compare, for example, a neighborhood that is economically, socially, or functionally heterogeneous with several neighborhoods that differ from one another but are internally homogeneous.

Students often seem to focus on visible trash or on the quality of a paint job on a house, calling that "bad design." Encourage the participants to look deeper into the way people use the buildings and the spaces between them.

These neighborhoods are much better comprehended if experienced at both pedestrian and automobile speeds.

Visitors from the Department of City Planning, architects, and city planners are excellent critics to have available while the teams are working on their neighborhood designs.

Individuals laying out a two-dimensional plan on paper tend to get caught up in the appearance and symmetry of the layout. They need to be reminded that their plan is only an abstraction of a three-dimensional environment used by people and constantly changing.

The conclusion of these learning experiences is an excellent time to check whether the participants have clearly focused their ideas for the optional long-term projects.

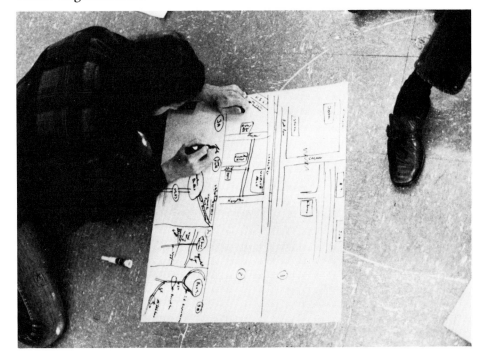

proposing a neighborhood

Ask the participants to form three to five person teams. Each team will draw a scheme for a neighborhood. "First indicate the activities you would like to see made possible for your residents. Then diagram the physical characteristics that will best support these activities. Annotate your sketches to help overcome the difficulties of communicating a multidimensional scheme in only two dimensions. Include the age range and interests of the inhabitants, for example, and places they will meet."

Have each team present its design to the whole group for analysis. Ask each team to explain all the ways their plan would be useful to inhabitants. Then ask the audience to evaluate how well the proposed physical layout would in fact function as the designers say it will. Now ask participants to describe the designing process actually used to arrive at the plan. Was it one person's influence or the rationale and ideas of the entire group that brought about a decision?

This neighborhood development proposed 614 dwellings of four types, including row houses. What differences do you see between this and a neighborhood planned on a grid of streets? What lifestyles would be enhanced in this physical setting? Which lifestyles would be adversely affected? What priorities of values have the designers expressed in this scheme. What trade-offs have been made to obtain the qualities you can identify from this plan?

VISUAL SITE PLAN

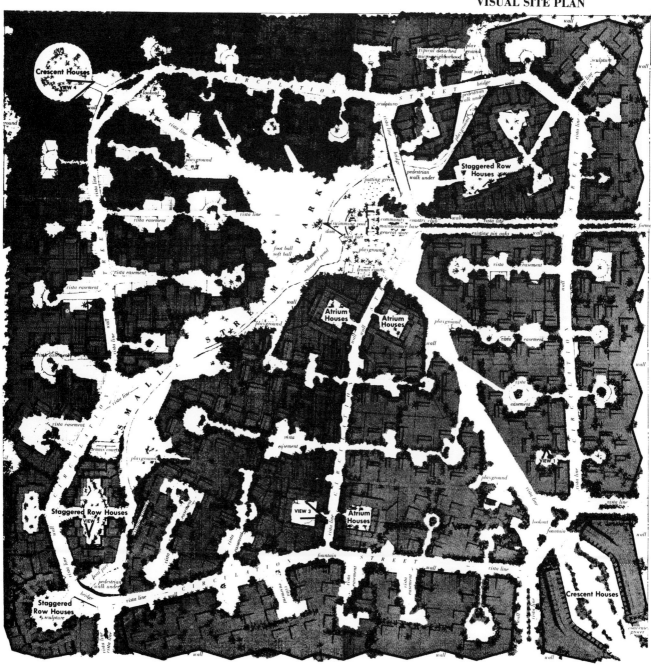

Identify one or more
neighborhoods and arrange
transportation to them, if
necessary.

copies of "neighborhood
questionnaire" handed out at the
end of the previous session

pictures and maps of the
neighborhood as it is now and as it
looked fifty to one hundred years
ago

white butcher paper

felt-tipped pens

further exploration

Ask the participants to sketch from memory the neighborhood they visited.
Then distribute current maps and aerial photographs of the area for comparison.
The participants should analyze the differences between their perception and
memory of the neighborhood and the way it actually is. "What are the
differences and how can they be accounted for?"

"Write an essay on how your life would be changed if you lived in the
neighborhood just visited or designed."

check list

How thoroughly has the participant integrated into his project the basic ideas
discussed during the neighborhood visit?

Did the participant in his presentation of an ideal neighborhood emphasize and
articulate well the people's use of the physical structures represented in the
layout?

resources

Jacobs, Jane: *The Death and Life of Great American Cities*, Vintage Books, New
York (1961). Several chapters pertain to neighborhoods and what makes
"people" places work. Highly recommended.

Lynch, K., and Rivkin, M.: "A Walk Around the Block," *Environmental
Psychology: Man and His Physical Setting,* Holt, Rinehart and Winston, New
York (1970) 631-642. A research project investigating how individuals perceive
their landscape, what impresses them most strongly, and how they react to the
landscape. Similar projects could be set up by the participants.

Norberg-Schulz, Christian: *Existence, Space, and Architecture*, Praeger, New
York (1971). See Pages 39-59 for definitions of place, path, domain, and district.

Pratt, James: "Neighborhoods: A Matter of Choice," *American Institute of
Architects Journal* (May 1970) 51-55. Designing neighborhoods as manageable
pieces of the city.

Spreiregen, Paul D.: *The Architecture of Towns and Cities*, McGraw Hill, New
York (1965). See Chapter 8, Residential Areas.

Zeisel, John: *Sociology and Architectural Design*, Russell Sage Foundation,
New York (1975). A discussion of how social science research can aid the design
process.

U.S. Geological Survey maps can be purchased. See Resources, Chapter 15.
Photographs and maps can usually be found in the city's library or planning
department and in the state and federal street and highway departments.

Public constraints

20

Limiting individual freedom to fulfill group needs: discussion and a project help to uncover the hidden trade-offs of environmental decision-making.

there are hidden controls on the individual

Often indirect controls are exerted on the few in response to the over-riding needs of the many. Many of the controls can be recognized because they are accomplished through the power of eminent domain. Examples are provisions for freeways, mass transit, and water, gas, telephone, and electric utilities.

Indirect controls also include attempts by the government to provide order and safety. Legislation codifies group standards for zoning, building construction, and fire prevention. If such legal guidelines are not frequently updated they can stifle innovative development. A test question is "Does a particular code create more problems than it solves?"

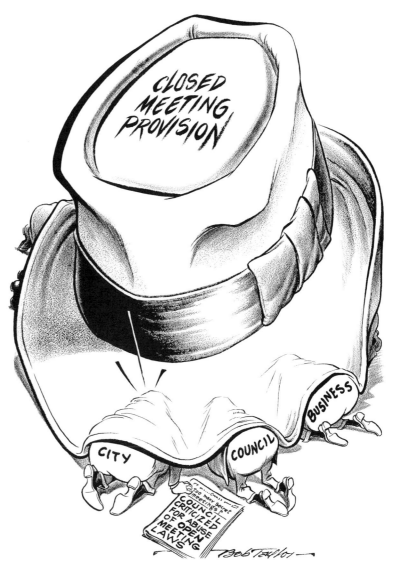

Hidden policy making limits our participation in government.

"Keep this under your hat, men . . . but they say we're spending too much time under here."

market place attitudes

Both federal and private lending agencies want to put their money in developments that are "certain" to sell because they are vernacular and familiar to the largest number of individuals — the potential customers who create the market. This policy reinforces fads and limits choices. Design is reduced to standardized formulas of room sizes. The houses are then sold on the basis of superficial details such as a Mansard roof designed to connote a French Provincial dream house rather than on the basis of their functional design and sound construction. This marketplace attitude indirectly controls creativity. A design for an innovative house is often difficult to finance, because until this design too has become accepted, it is considered a risky investment.

Land speculation is encouraged because the value added to the land by public investments in streets and utilities is not taxed. Our tax structure also allows builders to make quick profits by keeping their first costs low. After a few years the bloom is off the cheap materials; the builder has his profit and is gone. The community is left with an incipient slum, which still must be amortized for another fifteen, twenty-five, or even thirty-five years.

the hand of government

Taxation by the government can be used to enhance the quality of the environment or to penalize it. A fine old building can be torn down to obtain the lower tax rate assessed on undeveloped land. An expensive building contributing a plaza or other open area to the public may be tax-penalized for having done so. Thus, the Seagram Building in New York City has paid higher taxes than a less carefully built structure on a similar site.

Governmental decisions rely heavily on expected economic benefit to the community. The standard commonly used to evaluate alternative proposals is known as the cost/benefit ratio; that is, how many dollars will the project generate compared with how much it will cost. Cost effectiveness sounds like a hard-nosed criterion for governmental expenditures. Actually it is difficult to identify all factors, particularly the intangibles, that bear on even a simple issue, such as whether or not the city should build a day care center.

Furthermore, persons arguing pros and cons on an issue select those variables that bolster their own point of view. Some of the factors that are raised can be assigned only rough dollar estimates to plug into the cost/benefit formula. For example, how can we accurately calculate the benefit to the community of freeing mothers to take jobs? Even when costs can be estimated for items such as personnel and utilities, the estimates are based on the unknowable future. Personnel and utility costs may soar, causing the center to become a burden on future taxpayers.

the power of social mores

Society acts both directly and indirectly to control the actions of its members. The community sets the limits of how much deviation from the group norm is acceptable. For example, all the dwellings on one street usually have the same degree of communication with one another. The yards may all run together, or have visually transparent chain link fences, indicating a general "neighborliness." Or each yard may be enclosed by shrubbery, suggesting that these homeowners value their privacy. If one of the dwellers in an "open" block erects a brick wall, he is considered unfriendly.

The mores of the group may be conveyed directly, by laws, or symbolically. A subdivision can be restricted by deed to specific economic groups or house designs. Unspoken restrictions may discourage from moving into a neighborhood those persons whose tastes do not conform. Restrictions are motivated by fear of the unknown, such as what might decrease property values. Although the diversity in a heterogeneous neighborhood might enrich our lives, we tend automatically to shop for a home where the people are like us in their tastes, economic level, and family size.

Social mores may control when least expected. Architect Brock Eustice designed his house as two, two-story high cubes to solve the problem of building on a lot 45 feet wide, overlooking a highway. Neighbors who perceived the house as orange crates or a spite fence, reconstituted an architectural control committee which had never met, sued and won on the basis of a deed covenant. The bank withdrew further construction funds. On appeal, the state supreme court agreed that the house must be torn down. Mr. Eustice, nine years later, is paying off the mortgage on an unusable lot sold to him by a member of the architectural control committee.

the influence of special interest groups

Policies may be made by a small number of people who accrue power. Exerted directly or indirectly behind the scenes, this covert control can be an influential force. For example, a freeway may be located to favor special interests. Personal power can be obtained through commitment and by working more energetically than everyone else. Often money gravitates toward such energy. Thus personal power means access to money and to other influential people. A promising political candidate is heavily backed with campaign funds to supplement his own meager resources. Money sees the energy as a means to effect its own ends — whatever they may be, selfish or altruistic, limited or broad. Hidden decision-making is hard to document but may profoundly affect a citizen.

Spite Fence

After years of bickerings

Family one
Put up a spite fence
Against family two.

Cheek by cheek
They couldn't stand it.
The Maine village

Looked so peaceful.
We drove through yearly,
We didn't know.

Now if you drive through
You see the split wood,
Thin and shrill.

But who's who?
Who made it,
One side or the other?

Bad neighbors make
good fencers.

Richard Eberhart

smoking out better alternatives

The process of legislating, reviewing, and modifying public constraints is part of living democratically in a community. Some of the controls on our lives are visible to all; others are covert. In most cases those constraints are trade-offs. We give up one thing; we get another. We tend, however, to leave such decisions to our elected officials and their staffs. We don't often enough ask "Should this control be updated? Is there a better alternative?"

In discussing proposals to enhance the built environment, there is an additional constraint: the inherent difficulty of communicating alternatives. Most people are unable to visualize a three-dimensional building from a two-dimensional drawing. The effects that time and human interactions will have upon a particular environment are not expressed and therefore are minimized or ignored. All of these factors limit imaginative land use and building. Possible options need to be communicated clearly, so that attention can be directed to "How can we make the best alternative become a reality?"

identifying public constraints

"In previous exercises we have discussed the individual satisfying his personal needs. We also considered how we depend on our interaction with an external world of other people and our interaction with an extended physical environment.

"What constraints does the community place upon the individual who is attempting to build a personal environment such as his home or office?" List the participants' examples and help fill out categories of controls such as social mores and economic controls that may have been overlooked.

"Give specific instances of cases in which controls had an adverse effect on the environment." (The uniform setback on residential streets gives order to the street, but also could be considered monotonous. We need a balance between order and chaos.)

"What about cases in which the effect of codes was positive?" (stoplights, fire codes) "How can you as an individual have a say in establishing or modifying controls? Can we cite a case history of a local current issue?"

The proliferation of signs along thoroughfares has led to attempts by cities to regulate their size and quantity. Drafting a fair sign ordinance is difficult when the same sign is considered esthetically pleasing to some, but visually polluting to others.

our experience
Face to face contact and persistence are the best way to break through the bureaucracy of city staffs who are unlikely to volunteer information.

long-term projects

"Because constraints are so often hidden, it is not easy to identify them in any particular instance. For this reason each of you in teams of four or five are to take a current controversy and present it in your long-term projects as a case history.

"For example, Joe Doe, a developer, is trying to get a zoning change to build a shopping center adjoining a residential area. To identify where pressures are being exerted to control this particular piece of the environment, you might interview the developer, the head of the City Plan Department, members of any citizen group opposing or promoting the change, the City Traffic Engineers, newspaper reporters, and the bank financing the project.

"You should also collect all newspaper copy on your topic. This will help you identify the issues and some of the people behind them. You will have to be very persistent to get glimpses of behind-the-scenes maneuvering, since the persons making deals are not likely to reveal their stratagems. Reporters should be able to help you find the many different sources of pressure. In addition, be sure you have interviewed people on both sides of the issues; each group will know what the rumors are about the opposing side."

further exploration

"Play one of the urban and environmental games; for example, *Urban Dynamics*, which covers 'basic structures and interlocking systems in the growth and development of a metropolitan area,' in a five- to six-hour playing time. The complete kit for twelve to twenty players is available from Urbandyne, P.O. Box 134, Park Forest South, Illinois 60460.

"Join a local citizens group as a volunteer to help investigate problems and assemble data concerning city issues."

check list

Can the participant identify both the positive and negative aspects of the indirect controls placed upon him?

Can the participant articulate the possible roles for himself in the decision-making process of his city?

resources

Bacon, Edmund: *Design of Cities*, Penguin Books, New York (1976). See Pages 302-303 on design as process.

Bagley, M., Kroll, C., and Kristin, C.: *Aesthetics in Environmental Planning*, prepared for U.S. Environmental Protection Agency, Washington, D.C. (1973).

"A Cube House vs. the Squares," *Life* (Nov. 14, 1969) 83-86. Report of neighborhood lawsuit against architect Brock Eustice.

Gordon, James Stewart: "We're Poisoning Ourselves with Noise," *Reader's Digest* (Feb. 1970) 188.

Manley, R., and Fischer, T.: "The Effect of Aesthetic Considerations on the Validity of Zoning Ordinances; The Status of Aesthetic Land Use Controls in Ohio," The Cincinnati Institute, Cincinnati, Ohio (1974).

Planning and Community Appearance, Henry Fagin and Robert Weinberg (editors), Report of the Joint Committee on Design Control of the New York Chapters of the American Institute of Architects and the American Institute of Planners, Regional Plan Association, New York (1958). The impact of public boards, legislation, and court decisions on public esthetic controls.

Zuckerman, D. W., and Horn, R. E.: *The Guide to Simulation Games for Education and Training*, Information Resources, Inc., P.O. Box 417, Lexington, Mass. 02173. A review of over six hundred games.

Groups, such as the League of Women Voters and the Sierra Club, who have worked on land-use reports and urban problems often have printed materials for public distribution.

21

Movement systems

Differentiating movement patterns in the city: participants analyze their own use of existing movement systems as a basis for suggesting qualities that would enhance their function.

the ways we move need to be interrelated but differentiated

The number and variety of movement systems symbolize the dynamics of the second half of this century. Our expenditures for movement systems and the extent to which we use eminent domain to create them reflects the degree to which their construction takes precedence in our culture.

Freeways, transitways, streets, railways, sidewalks, bicycle paths — all are components of a city's movement system. Each component has patterns of movement across one or more of the others. The safest movement system must take into account not only the particular range of users, vehicles, and speeds, but also must provide carefully planned transitions between these components. Fast-moving cars do not mix with slow-moving pedestrians.

Movement corridors of all kinds need to be made more comfortable for their users. Landscaping needs to be chosen to screen and muffle visual and noise pollution, to make it both safer and more pleasant to drive. To encourage use of rapid transit there must be adequate parking for suburban commuters, sheltered seats at stations, and covered walkways at interchanges between modes of transportation.

Cities continue to sprawl; but energy, time, and resources become increasingly precious. Should the city encourage that sprawl by spending more and more money to accommodate vast numbers of people moving between home and work in private cars? As freeways are built they attract commercial development that rapidly generates new congestion. To localize job opportunities and services might in the long run cost the city less and allow citizens to work, shop, and find recreation nearer home. It is interesting to speculate: What kind of urban movement systems will we have in the future? How much will the current investment in conventional systems dictate those systems of the future?

Movement lines first drawn on the prairie by horses and wagons became the streets *foreground, left to right.* Two railroads that crossed where the construction stops have come and gone. The freeway cuts a new swath above the old patterns.

living in your car

Explore in discussion to what degree we in the United States are a car-centered society. "In what ways is our daily life shaped by the relationship with our car? What is the highest total number of hours any of you has spent in a car since you started your logs? How many hours have you each averaged per day? At that rate, if you live to be eighty years old, what total number of hours will you have spent in your car?"

In the abstract, a movement system is what transfers us from one place to another. In our lives, the car together with our driveways, streets, freeways, and parking lots is the movement system we most often use. We have invested a great proportion of our natural and human resources in the car and its movement systems, including much of our urban living space. "How many alternative transportation modes between home and work are available to you? Which ones might you try?

"What is it like to spend so much of your time in the car? Why are people so resistant to new modes of travel? Do we fear to lose the sheer pleasure of being alone and in control of a part of our lives? Some people describe a gut feeling of independence from the mere physical contact of their hands on the steering wheel as they drive along. The decision of which way to go and how fast is theirs.

"What makes a car trip pleasurable or unenjoyable? Does the design of the road make any difference? What about traffic and directional signs? Are there too many signs to read in time to make a decision? Have you noticed in what ways landscaping can make a difference in your car travel? Do you recall from the first half of the book your experiments with the perception of objects as a function of your rate of speed? For example, what differences do you observe in the rhythm of building forms, texture, colors, sounds, and smells while circling downtown looking for a parking place and later when you see the same movement corridors on foot? Try to summarize the important factors to consider in designing any movement system from the point of view of experiencing it at different speeds. Include, for example, travel in the air, on a freeway, on a country lane, on an urban street."

moving in the suburbs

The car is given priority in our society today. The freeways and "mix-masters" necessary for great numbers of fast-moving vehicles dominate our cities.

How successfully a freeway functions depends not only on how fast the traffic can move but how well it is interrelated with the slower moving pedestrians, bicycles, and cars on adjoining neighborhood streets.

Note that the creek has been channeled under the mix-master.

interconnecting the corridors

"Using the tracing paper and site maps, outline the principal movement corridors criss-crossing the city. Mark the neighborhood's connection with the city at large. Is the overall pattern of major movement systems clear enough to be memorable? How many connections are there with the neighborhood studied? What are the alternatives available to the residents for going downtown? across town? Which of the routes would provide the most enjoyable experience for residents of all age groups?

"How can the residents get to the nearest public swimming pool? How can they reach an interstate highway? How do they reach the airport? What choices of transportation do they have?

"The patterns already drawn on the tracing paper represent the primary movement systems, that is, the network designed for high speed and large numbers of vehicles. Are there gaps in this network, or can one travel about or through the city without leaving this movement system?

"Now mark in another color the secondary car systems, showing at what points they feed into the primary, high-speed system. Now add the systems provided for low-speed movement, bicycling, and walking.

"Next, look at the blue lines on the city map that denote water. Is there a large lake or river in the city? Are there interconnecting smaller waterways? (These are sometimes also used as drainage ways.) Do they appear to form an interconnecting network that could also be a movement system? How are the intersections of these waterways with other movement systems handled? Are there any cities you can think of that use water, not as their primary system, but for some transit?

"What kinds of mass transit does the city have? What new types are being considered? What pattern of routes does the city's mass transit system have? (radial, a grid . . .?) Are the routes well designed to link where the people are to where they want to go?"

moving within the downtown

Four levels separate the parts of the movement system that interconnects within this shopping complex. The lowest level ties into rapid transit. The second level opens onto city sidewalks and streets. At the third level, pedestrian walkways bridge streets and lead into the second floors of two department stores. The top level connects directly with a parking garage.

Open entrances to the subway, visible from all shopping levels, help make public transit a more attractive alternative to the private car.

moving around in
your neighborhood

"Using your maps and your memory of the neighborhood you studied, analyze the different movement systems found in the neighborhood. While discussing each system, point out changes that would facilitate the use of that system by the local residents. Analyze the quality of each experience such as walking, bicycling, or driving a car.

"What different levels of movement did you find in this neighborhood? Where can small children go? Do you think children 'see' movement systems in patterns different from those seen by adults? (Isn't cutting through a yard part of a movement system for a child?) Where can the elderly go? commuters? service people?

"Analyze the various kinds of systems available in the neighborhood for pedestrians, bicyclists, joggers, trucks. How are the intersections of different movement systems handled? (streets, sidewalks, alleys) Are there any cul-de-sacs, dead-end streets, or loop backs? What positive or negative effects does this have on movement? For whom?" (Establish the difference between 'going through' and 'going to.') "What techniques are used to slow down auto traffic? What additional ones can you think of?" (tree-lined curves, hills, modulation of median width . . .)

"Are the alleys used as a secondary movement system? By whom? Is it a safe system for its users, for example, bicyclists? Is it a pleasant system for its users?"

Above, a low density conventional suburban neighborhood is organized for the driver of a car. Sidewalks abutting the curb are less safe for children and make impossible a stroll along the street at a comfortable distance from parked and moving cars.
Below, a new-town-neighborhood routes pedestrian movement parallel to the creek and through a park, rather than contiguous to streets. Note the provision for visitor parking in the cul-de-sacs.

TIMBER CREEK PARK

proposing a movement system

"Consider that your city has decided to begin or to augment a mass transit system that will connect the downtown business district, the airport, and residential and commercial centers. Working with your long-term project teams, decide what kind of system to use — past, present, or science fiction models. Using your map and tracing paper, propose a route from the commercial center of the neighborhood you have been studying to downtown and to the airport. Locate a neighborhood station for the new movement system and show secondary-transit (auto, bicycle, and pedestrian) connections that go to it from within the neighborhood.

"When the sketches are finished, examine each of the proposed schemes from the point of view not of a designer, but of a taxpaying citizen. Subject each project to the same kind of detailed analysis we have been making. What would be the effects of a given scheme on your life? Will it be safe, pleasurable, and convenient? Will it be a source of visual, noise, or odor pollution? Would you prefer that the primary system stop near your residence, or that a feeder system stop there? What will be the long-range result of each action? (Pressure for higher and higher densities of dwellings and commercial spaces occurs near primary-system mass transit stations.) What trade-offs do you personally prefer? How well does each proposed scheme function as one of the interlocking systems that serve the whole city? Remember, a surburban low-density area cannot economically support mass transit without large subsidies. Will the benefits from the new system outweigh the expected costs?"

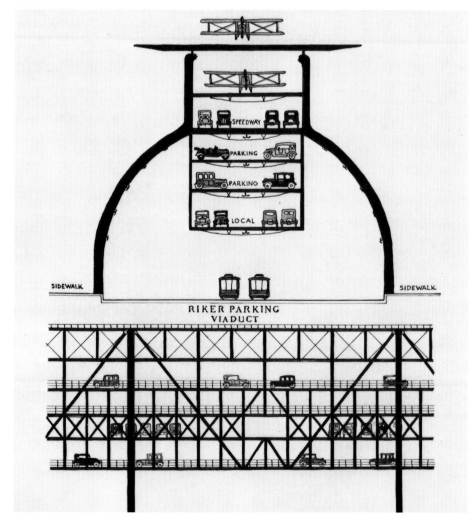

This whimsical movement system was seriously proposed to solve traffic congestion in 1925. The designers neglected to ask how their scheme would affect the surroundings. The mammouth steel structure would have blocked the view from the White House as it helped vehicles speed from the Capitol to Key Bridge along the nation's chief ceremonial street, Pennsylvania Avenue.

preparation

white butcher paper

tracing paper

colored pens or pencils

time logs assigned in Chapter 18,
Personal Needs

maps of the city, past and present,
drawn to the same scale, if
possible

further exploration

"To experience in an exaggerated way the sensory impact of airplanes and cars, sit as close as feasible to the end of a runway or next to a freeway. Record the physical, visual, aural, and olfactory impressions you receive."

check list

Can the individual sketch the principal patterns of routes that compose the movement systems in his city?

Has the individual demonstrated in his project various qualities that would be desirable in movement systems?

resources

Bacon, Edmund: *Design of Cities*, Penguin Books, New York (1976). See Pages 252-262, 317, and 322 on movement systems.

Dettelback, Cynthia Golomb: "In the Driver's Seat: A Study of the Automobile in American Literature and Popular Culture," Case Western Reserve University Dissertation (1974).

Halprin, Lawrence: *Freeways*, Reinhold, New York (1966). An imaginative look at what the freeway has done to us and what to do about it.

Heinlein, Robert A.: "The Roads Must Roll," in *Science Fiction Hall of Fame*, Robert Silverberg (editor), Avon Books, New York (1970). Why not have roads move, instead of the cars? A story that dramatizes society's total dependence on a particular technology.

Jacobs, Jane: *The Death and Life of Great American Cities*, Vintage Books, New York (1961). Chapters 2, 3, and 4 discuss "The Uses of Sidewalks."

Lynch, Kevin: *The Image of the City*, The MIT Press, Cambridge, Mass. (1960). Chapters 3 and 4 discuss the use and design of movement systems, including paths.

Macaulay, David: *Underground*, Houghton Mifflin, Boston (1976). Fascinating line drawings reveal the way goods and services move underground.

Rapuano, Michael, *et al.*: *The Freeway in the City, Principles of Planning and Design*, A Report to the Secretary, Department of Transportation, by the Urban Advisors to the Federal Highway Administrator. U.S. Government Printing Office, Washington, D.C. (1968). A clear, balanced exposition of urban and rural freeway design that includes many factors in addition to provision for speed. A good glossary defines unfamiliar terminology.

Rudofsky, Bernard: *Streets Are for People*, Doubleday, New York (1964). An enjoyable history of streets as the lifeline of urban civilization and the role streets play within a community.

Orienting in the city

22

Orienting in the larger urban environment: a field trip
enables us to study the size and complexity of the city
as it affects us and we effect changes in the city.

how do i get to...

The size and complexity of cities have a tendency to
overwhelm and alienate us. At the same time, by the
very diversity they offer, cities have the possibility of
stimulating us.

We can feel at ease in the larger urban environment if
we are able to perceive patterns by which we can orient
ourselves. Because this is a basic need wherever we are,
orienting devices are vital in the design of a city.
Without a means for orienting, we feel lost and
confused. Then we tend not to make full use of the
city.

We perceive by contrast, therefore we cannot orient if
everything looks the same. We have to find differences
to know where we are. The variations that attract
attention differ according to who is looking at them.
One person identifies a particular row house by the
sycamore tree in front; another, by its distinguishing
blue door. Tall structures, because they are so visible
and easily remembered, are used for orientation. At the
same time they determine the character of a city and
give it architectural variety.

Some orienting devices are built to orient, such as signs
and university towers. Others exist naturally, such as
rivers. Church towers built as symbols, water towers
built for functional purposes, and office towers built
for economy, all serve also as orienting devices. We
refer to them to give directions, at times, even after
they have been torn down. On a smaller scale, the
school in the subdivision or the gas station on the
corner stand out from blocks of similar houses.

We also rely on the physical environment to orient
ourselves in time. Events in our lives are often
pinpointed or dated by the construction or loss of
familiar landmarks.

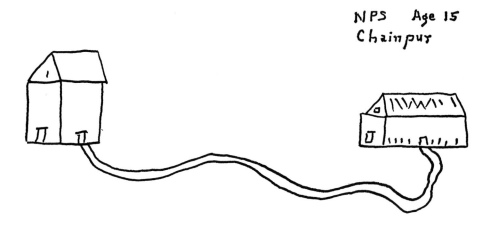

JD Age 11
Honolulu

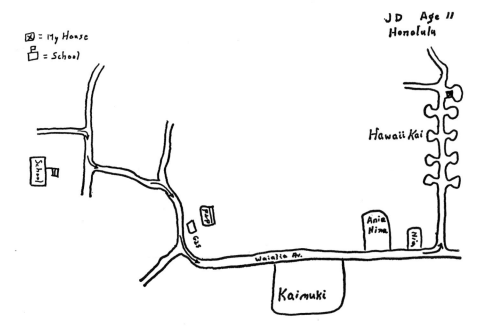

⊠ = My House
🏠 = School

School

Park

Gas

Waialie Av.

Kaimuki

Hawaii Kai

Anie
Nini

Niu

orienting along movement systems

The way we orient is partly determined by our culture. The Nepalese boy from Chainpur shows the way home from school as a *process of going* from one to the other. He has not represented spatial relationships; in fact, the house and the school are not on the same street or path.

The American child from Honolulu has represented his home and school as abstract symbols, not pictures, and has drawn numerous spatial and directional clues to map his route home.

Ask the participants to choose partners. Ask each member of the pair to give his partner oral directions on how to get to his home from their present location. The partner should write down all details of the directions he is given.

"Do the directions you were given form a complete pattern, like a city map, or do they form a linear route? Do you visualize a bird's eye view of the area and then mentally trace a route? Or do you visualize a route on roads you've previously taken? Which directions are easier to remember, a linear route, or a route built into a more complete description of the area? What does this imply with regard to the planning of a city?

"Were there any orienting devices mentioned in the directions? What types of orienting devices were they?" (signs, railroad tracks, specific buildings)

"Can a city have too rigid an over-all pattern (as small towns often do) and therefore be monotonous? What types of patterns make orientation very difficult?" (Radial patterns of streets as in Paris or Washington are particularly confusing.)

comparing ways to orient

Ask the group to try an experiment to determine what kind of information enables one to get from one point to another the easiest. Give each group of four or five participants a different "map" to go from the starting point to an unknown designated meeting place. Each map should have a different type of directions. The "maps" could be, for example:

1   A scale drawing of a linear route containing no other information
2   A map consisting only of orienting devices such as towers, railroad tracks, the sound of a carillon, the odor of a bakery . . .
3   A standard map with written directions
4   Written directions based on the cardinal points ("5 blocks N, 20 yards E, 100 yards SE")

When all of the groups have arrived at the designated meeting place, discuss which "map" enabled the fastest arrival and why.

bird's eye viewing

Take the group to a high point overlooking downtown such as a public observation deck.

"What are the orienting devices that identify particular parts of the city for you from up here? Are they the same as those you use at street level? Can you find any of the neighborhoods we've talked about or visited? What pinpoints them from here?

"Are all the orienting devices buildings? What about freeways, large areas of trees indicating parks, and creeks? Can you identify any non-visual orienting devices from this high up?" (bells, chimes, noon whistles, odors)

"Using your maps, locate the various parts of the city we have previously discussed. What orienting devices helped you relocate each of these places from this bird's eye view? Name the street level orienting devices you use which differ from what you perceive as orienting devices while you are on top of this building.

"What distribution of land uses do you see? What boundaries define these uses? Mark on your map major changes in land use patterns and appearance that seem to have taken place during the past ten to twenty years. Save these maps. We will use them later in Dealing with Change."

orienting in center city

Reassemble the participants on the ground floor of the downtown building you are in. Assign teams of three or four participants three buildings or places to walk to and locate on their maps, meeting back at this location at a specified time.

"How did you locate your buildings? Which orienting devices helped you the most? Which were of secondary help? Were any of the orienting devices you used non-visual?" Ask each team to give oral directions for reaching the buildings they located. "Did any team give directions that did not rely on street identification signs?"

improving orientation

"Communicate to the group how you could improve orientation and add variety to or simplify a chosen part of your city by the careful placement of orienting devices. Use ways other than directional signs."

further exploration

"Compare maps of a number of cities, such as Los Angeles, Boston, Rome, Paris, Washington, London, New York, Tokyo, and San Francisco. Analyze from these two-dimensional representations which city appears the easiest to orient oneself in; which city seems the most interesting; which city shows the strongest or most complete over-all plan.

"In what ways does each city favor the pedestrian? the automobile? What types of movement systems will cities be using in fifty years? Select three of the above cities and suggest how each might physically accommodate to those systems."

check list

Has the individual analyzed what he uses to orient himself in the city?

Has the individual demonstrated in his project the ability to suggest memorable orienting devices that identify parts of the city?

resources

Bacon, Edmund: *Design of Cities*, Penguin Books, New York (1976). See Page 301 on the total organism, and Pages 302-303 for a clear exposition of design as process applied to city planning.

Cities: A Scientific American Book, Alfred A. Knopf, New York (1966). The diversity of new forms of human settlement.

Cullen, Gordon: *Townscape*, Reinhold, New York (1961). An analysis of how we are affected by the placement, color, texture, and scale of buildings as we move through the city.

Halprin, Lawrence: *Cities*, Reinhold, New York (1963). Abundantly illustrated examples of what cities should be like to satisfy man's personal, physical, and psychological needs.

Halprin, L., and Burris, J.: *Taking Part*, The MIT Press, Cambridge, Mass. (1975). A workshop approach to collective creativity.

Jacobs, Jane: *The Death and Life of Great American Cities*, Vintage Books, New York (1961). The impacts of traditional city planning techniques on the elements of the city that give it life, spirit, and character. A discussion of the dangers of change.

Lynch, Kevin: *The Image of the City*, The MIT Press, Cambridge, Mass. (1960). Includes discussions and maps of Los Angeles and Boston.

Pirenne, Henri: *Medieval Cities*, Doubleday, New York (1956). A synthetic approach to the economic awakening and birth of modern Western European urban civilization by a noted French historian and scholar.

Rasmussen, Steen Eiler: *Towns and Buildings*, Harvard University Press, Cambridge, Mass. (1951). Buildings as they have been used to shape cities.

Wurman, Richard: *Making the City Observable*, The MIT Press, Cambridge, Mass. (1971). Imaginative techniques for organizing complex data about cities into perceivable patterns.

Public and private values

Choice in building the environment: simulating a meeting of the city council to air the pros and cons of flood-plain management can reveal the conflicts involved in reaching environmentally sound decisions.

to get a balance: trade-offs

Many forces influence the physical and social forms of cities. Examples of political, social, and ecological determinants are existing movement systems, established building patterns, utility services, and public and private economic pressures. These forces interact continuously in every community. To make planning effective, it is necessary to be able to identify these forces and to know how to work with each.

A planning framework that responds *simultaneously* to private and public needs can effect orderly development, with amenities for both the individual and the group. The individual, in making sacrifices for the "public good," may himself derive direct benefits. The lack of a clear delineation between the public and the private good can be a source of conflict. What seem to be solutions may be transitory fads: at one time freeways and minibuses; at another, high-speed transit. A plan allows us to put into context the long-range implications of daily compromise decisions. However, a plan becomes a strait jacket if it is not periodically and regularly updated.

Short-sighted decisions can be minimized if the long-range implications of development are considered. For example, a street can be routed through park land saving land costs now. But it will be more costly to buy equivalent park land when we need it for a future, denser population.

137

taking responsibility, making decisions

Ask the participants to imagine themselves as elected officials, urban planners, and citizens in a city hall drama in Hobroke, population 750,000. (See Map A.) The actors play the roles of the city council, city staff, consulting engineers, and private citizens who wish to argue for and against the issues presented. The decision to be made is whether or not the city council should allow the removal from the flood plain of a piece of private property adjacent to the Little Caravan Creek. Removal is accomplished by allowing the owner to fill that property with dirt until its elevation is greater than the city's established one-hundred-year flood line. After the filling, the property is considered to be no longer in the flood plain. The owner may then build on the land.

The case going before the Hobroke city council raises two issues. First, where do you draw the line to balance the rights of the private owner with the welfare of the community? This is one argument: "Persons should be able to do anything they wish with their private property. It is a basic civil right." At the other extreme is this argument: "The welfare of the community is more important than one person's civil rights."

A second major issue arises because the city is held by the courts accountable for flooding: What is the best way to manage or control flowing streams and their attendant flood plains? Again, there are two extreme positions: "Don't build in the flood plain, then nobody gets flooded"; and "Man has the technology to protect himself and his property and the right and the duty to do it."

Of course, between the extremes of either issue there are dozens of positions, all backed by sound rationale. And the position a person finds himself in at any one time may be very different from his position the next time he faces the issue. It takes considerable thought and experience to recognize the controlling variables in a situation. It requires time to decide (1) the relative importance of each variable and (2) what will be the possible consequences of a decision. In a particular instance, such as filling in Little Caravan Creek's flood plain, two highly valued "rights" may be in conflict.

Ask the participant to study the maps and to consider the ramifications of filling the marked site. Encourage them to question the different positions. This is an appropriate time for professionals to join the group and add their expertise. There will not be answers to all the questions. However, city council members often must make such decisions on the basis of incomplete knowledge and misinformation. It is important that the participants think the problems through carefully and assign relative values to those variables they consider important.

After the discussion ask each participant to select the character he will portray in the drama and the position he will take on the issue.

our experience

Some groups get so involved that they demand very technical information. Visiting experts add a great deal to this exercise. If allowed to, the participants will argue the entire session. Therefore, set a time limit within which the council must vote.

Mr. Hayes is the property owner. The property is adjacent to the Little Caravan, within the presently designated flood plain as defined by the Department of Public Works, along a tributary feeding into the Caravan River mainstream. (See Maps A and B.) Mr. Hayes is requesting permission to fill his property with dirt beginning at the 445-foot contour until it is level with the remainder of his land at the 451-foot contour. (See Map B, "site for proposed fill.") The entire property will then be one foot above the one-hundred-year flood line as determined statistically by the Department of Public Works. Mr. Hayes has already paid his loan commitment fee for development, and under the terms of the loan he must begin construction in sixty days. The council, to develop revenues for an ambitious capital improvement program, has recently raised taxes on undeveloped land, forcing Mr. Hayes from his position of holding this land, which has been in his family for a long time. He is nevertheless public spirited and has offered the city at his cost one acre of land for the expansion of a nearby day care center.

Mrs. Clements is the city's Mayor. There are six or eight additional council members who will listen to the arguments of all individuals who are speaking to the issue and may ask questions at the Mayor's discretion. The Mayor will direct Mr. Hayes's hearing before the council, decide who may speak, when, and for how long. The council members must come to a decision and vote within a designated time. The Mayor votes only in case of a tie. The council must either grant or deny at this hearing, permission to fill the land. If they deny Mr. Hayes permission to fill, they may do so "with prejudice," which means he must wait three years before returning with another request to fill or they may deny the case "without prejudice" which means Mr. Hayes may return at any time with a revised request to fill.

Mr. McKay, Director of Public Works, is responsible according to the city's charter for "flood prevention and protection of lives and property from flooding" within the corporate limits of Hobroke. Building permits are issued by a section within this agency. Mr. McKay should write a letter delivered to each council member stating his reasons for *approving* the fill permit. He should present among other reasons, his department's existing plans to channel a wider stream bed south of Hopie Lane and to concrete the bottom section to facilitate faster discharge of flood waters, thus reducing the volume of water at any given point for any length of time. The channel is designed to join the already channeled stream Southwest of Spruce Boulevard.

Mrs. Frost, Director of Parks and Recreation, is responsible for acquiring and maintaining parks, recreational facilities, and open space. She sends a letter to each council member saying among other possible things that the city maintains park and recreational facilities at Hopie Lane and Caravan Street and at Caroline Highway and Caravan Street three miles south of the property at issue. Although she has had some discussion with area homeowners wishing more park space, her department has no immediate interest in this case. The area is on no development plan for parks.

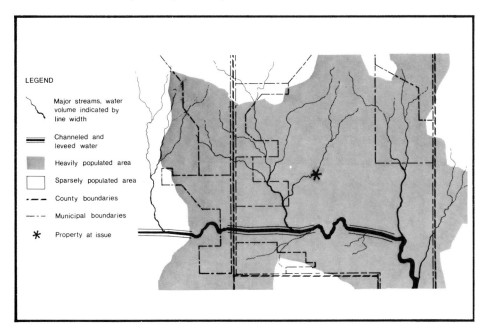

Map A

Mr. Holden is President of the Caravan Estates Homeowners Association. This information group of private citizens consists of two hundred dues-paying families living in the immediate area and an additional one hundred and thirty-five families included on the mailing list. For two years the group has been in communication with the Park Department, the Department of Urban Planning, and the council member from their district about turning Little Caravan Creek, beginning at the railroad and extending north, and its immediate flood plain, into a "linear park" or greenbelt. They oppose Mr. Hayes's request.

Miss Carraway is the Director of Urban Planning. Her department is responsible for overseeing development of the city as a whole so that the city functions well for all its citizens. Her staff is divided and will submit two reports to council members concerning this issue. Some of the staff think a linear park or greenbelt on the Little Caravan Creek would stabilize a neighborhood threatened by increasing small commercial operations. They predict deterioration of home and property values and therefore increasing urban rehabilitation if the neighborhood is not stabilized. The rest of the staff thinks the city should let the "invisible hand" of economics decide what happens to the area.

Mr. Sikes is Chairman of Citizens for Lower Taxes. This citizens' group whose support base is city-wide, objects to the fill request. Mr. Sikes maintains that the only reason the Department of Public Works approved the request is that they have already drawn extensive plans for a channel to increase flood water velocity and thus hasten its discharge during the big floods. They say the floods will statistically come only once every one hundred years, but the statistics are based on only thirty-seven years of data, and during that period there has already been one such flood. Further, Mr. Sikes maintains that channeling means spending thousands of dollars to subsidize a few landowners who wish to build in the flood plain. He contends that if fill in the flood plain were not permitted, there would be no flooding and no need to spend taxpayers' money for the channeling.

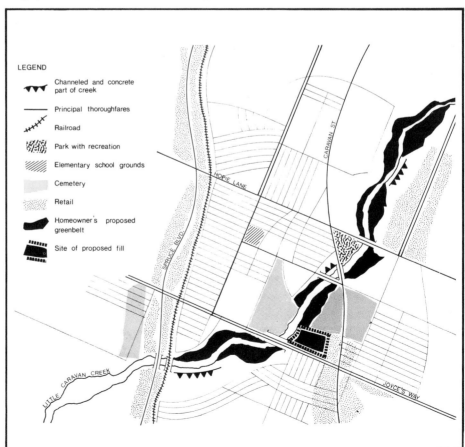

Map B

Miss Larry is the ten-year president of both city and regional ornithological societies. The sightings of purple-throated greebies have shown an alarming decline for the past eight years. They have just been placed on the endangered species list. A principal nesting ground is the heavy forest cover in this immediate flood plain. Miss Larry has evidence that the changes proposed will not only do away with the birds' nesting grounds, but will also do away with the greebie weed in the downstream backwaters of the flood area. The greebie weed is a primary source of food for the birds and a decorative winter plant used by the community. This habitat is the only one remaining in the county. In this area there are numerous two-hundred-year-old water oaks, also rare in the region. Miss Larry's constituency is against the linear park because it, as well as development, would destroy a rich and handsome ecological site. Her opponents see it as just snake-infested weeds.

Once the participants have prepared their chosen roles, they must lobby their positions with members of the city council. Mayor Clements may then call the council into session and begin the formal hearing.

analyzing the drama

Ask the participants to detach themselves from the drama in which they just participated and analyze both the decision-making process and the specific implications of the adopted solution.

"In your opinion was the council influenced more by the logic of the arguments or by the status and personality of the participant presenting them? Whose opinions had the most influence and why? Which arguments were short-range answers? Which were long-range? A solution is never a problem-free situation; it is a new direction. What specific problems do you foresee that will eventually arise and need to be dealt with in the future?"

Compare the way two creeks in one urban environment were handled, here, and on page 137. Originally the two creeks were very similar.

preparation

copies of Maps A and B

list of local groups and agencies
corresponding with the ones given
roles in the learning experiences
(optional)

further exploration

"Visit public hearings, city council, and planning commission meetings. Try to identify specific cases where each group draws the line between public and private values."

check list

How effectively did each participant present and defend the role he took?

Did some of the discussion and debate become heated?

During the discussion, did the participant attempt to articulate his philosophy of private vs. public values?

resources

Ball, M., Bialas, W., and Loucks, D.: "Structural Flood Control Planning," *Water Resources Research* (February 1978).

Bialas, W., and Loucks, D.: "Non-structural Flood Plain Planning," *Water Resources Research* (February 1978).

James, L., Denke, A., and Ragsdale, H.: "Integrating Ecological and Social Considerations into Urban Flood Control Programs," *Water Resources Research* (April 1978), American Geophysical Union, 1909 K Street N. W., Washington, D.C. 20006.

Open Space, Identifying the Critical Areas, prepared by the North Central Texas Council of Governments, Arlington, Tex. 76011 (1975). A resource to help lay people learn the terminology and techniques that will enable them to evaluate the natural features of land and to participate in the planning process.

"Residential Storm Water Management: Objectives, Principles and Design Considerations," published jointly by the Urban Land Institute, the National Association of Home Builders, and the American Society of Civil Engineers (1975). L.C. 75-34759. Available from the Urban Land Institute, 1200 18th Street N.W., Washington, D.C. 20036. Description of flooding problems and alternatives for handling them.

Public spaces

24

The impact of public spaces: a field trip reveals how the design of publicly used environments enhances or impedes their functioning.

public spaces are community living rooms

Places used by the community, whether they are privately or publicly owned, are considered "public spaces." The types and sizes of those environments are as varied as the groups using them. They range in size and function from a telephone booth containing one person, to a bus carrying thirty, or to airports and shopping centers serving hundreds. They are those squares, malls, plazas, arcades, and galleries where people come together to exchange information, buy, sell, recreate, rest, and participate in rituals. Public spaces also include movement corridors of all kinds, and parking lots.

The qualities of public spaces are as critical to our lives as those of private spaces because we spend so much of our time in them. Open spaces, including streets as well as plazas are *entities* that have all the qualities of interior spaces. The volume and shape may be defined by the exterior walls of surrounding buildings or by natural vegetation. Texture, color, pattern, and other qualities contribute to the whole, whether the materials are natural or built, or some combination of both.

Our sense of well-being is affected by the qualities we encounter in public spaces. Spaces have an "atmosphere" that determines largely how they are responded to and used. Underground walkways are often avoided because of

their forbidding atmosphere. Dampness, darkness, and the repelling texture of tile or concrete suggest danger. Fountains invite adults to toss in coins and children to retrieve them. Avenues with tree-shaded benches invite pedestrians to linger and rest.

Publicly used environments are not always designed primarily for the people who use them. Provisions for cars, or for maximum financial return on surrounding property can have priority over how well the public space meets people's needs. Such priorities reflect an individual's or a group's value judgments and do not necessarily represent the only solution to the problem. The priorities set up by the builders of Dulles Airport, Washington, D.C., contrast with those set up by the builders of O'Hare Airport, and therefore the solutions to these similar problems differ. Office buildings are covered with mirror glass for economy and to create an effective sculptural image. The fact that their smooth high walls sometimes create an unpleasant downdraft on the vulnerable pedestrians below may not have been considered.

Public spaces not only facilitate and enhance our private, daily lives, they also say symbolically what we think of ourselves. We gain identity as a community through our public spaces for they, as much as our individual buildings and activities, represent us to the rest of the world.

Public environments are a critical component in the design of the city because they are part of the over-all matrix of the city. They provide their surrounding areas a contrast in scale, materials, and architectural forms and they are the primary containers for our public activities. Public spaces are dominant, visible nodes in the fabric of the city, providing orientation and variety for us as we move through it.

The series of four views illustrates different intended uses of public spaces. On the previous page, the Faneuil Hall area in Boston has historically been a center of diverse activities. Shops, restaurants, and shaded outdoor seating have updated the area for contemporary uses and have restored some of the 19th century charm.

In contrast, the Quadrangle, pictured on this page, was built as a small center for impulse shopping and open-air strolling and eating. Access is provided for cars but they are carefully separated around the periphery.

Identify other functions you think were intended in these spaces. What additional activities could take place in them?

public spaces: their importance, their design

Ask each participant to add up the time he has spent in public spaces (other than streets and highways) during the last several weeks. Have each participant calculate, on the basis of these raw data, the approximate time he or she spends in public environments during a year. "What percentage of your time is spent in these spaces? How many hours does the entire group use public spaces in a year?"

Discuss with the group what characterizes a "public space." Help define public spaces by discussing them in terms of a continuum of spaces of increasing size and degree of public use. "As the population becomes more dense, will people turn inwards toward their private environments or outwards and use public spaces? What will effect this trend?"

Ask the group to select a large public space for a field trip. Their trip will be more helpful to them if they select a site that includes both positive and negative characteristics. They should include both interior and exterior spaces having qualities distinguished enough to command attention. Comparing two public spaces, if time permits, will help make the concepts more visible.

how public spaces work

As part of their long-term projects, the participants will communicate after the field trip the character of the public spaces they have visited. Before the trip, prepare them by reviewing methods of portraying and communicating a public space. "What means of communication would best convey the characteristics of a public space to a blind or deaf person? to a developer? to a florist who is a prospective tenant? to an architectural jury?" Suggest drawing a two-dimensional plan, elevations, and perspectives. They could build a model, take photographs, record sounds, odors, and textures, or write prose and poetry. Ask the participants to review the Public Spaces Questionnaire carefully before beginning the field trip and to anticipate how they will communicate through their chosen media the ideas the questions suggest.

Assemble at the field-trip sites chosen by the group. "Using the questionnaire, study the interaction of natural and built components. Then analyze these public spaces in terms of all the variables previously studied: light, materials, structure, texture, movement, both natural and man-made sounds, odors, scale, proportion."

At the end of the field trip, but while still at a site, ask the participants to divide into small groups and exchange their initial thoughts on how to communicate what they have learned. They should attempt to grasp each other's perceptions of how to communicate in media other than the one they have chosen for their long-term projects. "After listening to others' methods, what limitations, if any, do you foresee in carrying out your intentions through your own chosen media? Did you select slides because you like to give slide shows, or because you considered it the best medium for expressing your ideas?

"Apply to your own project your analysis of the space we have visited; that is, using the techniques, medium, target group, and purpose you have already chosen, show others how you view this public space. Your assignment is to communicate, 'public space.' "

public spaces questionnaire

1   On a continuum of environments ranging from "untouched" to "built,"
    where would you place this site? Explain why.

2   Public spaces come in many sizes and some contain a great deal more
    activity and noise than others. Are the scale and level of activity in this
    space appropriate or inappropriate to the surrounding area? (The scale and
    level of activity of a municipal airport is obviously inappropriate if the
    airport is next to a residential neighborhood.)

3   The boundaries of a space are also the interface of the space and its
    surroundings. Are the boundaries of this space clear so that the space is
    well defined? Does the interface provide a pleasant and effective transition
    between the space and its surroundings? Do the entrances to the space
    contribute pleasure to arriving in it?

4   Movement into, through, and out of the space is critical to its successful
    functioning. Usually there are many scales and levels of movement —
    delivery trucks, garbage disposal vehicles, automobiles, bicycles,
    pedestrians. Have the corridors for movement within the space been
    arranged appropriately? Do the activities on these corridors interfere with
    the different kinds of users? Do they interfere with the intended over-all
    function and visual qualities of the space?

5   What efforts have been made to deal with large numbers of automobiles,
    both moving and parked?

6   How is protection against crime facilitated or hindered by the design of
    the space?

7   How does the space lend itself to alternative types of activities? Is the site
    arranged so as to limit its uses? Would you feel comfortable in this space
    alone? for how long? with ten people? with a hundred people? In each
    instance, explain.

8   Is the site subdivided into smaller exterior and interior public spaces? Are
     the sizes and shapes appropriate for their designated functions such as
     shops, eating facilities, viewing places, or waiting areas?

9   What adjectives would you use to describe the atmosphere of this place?
     What physical components exert the greatest visual impact on you?

10   Was this public place designed as a whole, or did it evolve part by part
     over a period of time? Why do you think so?

11   Under what priorities was this space designed? (For example, if it is a
     shopping center, was it the cars, the delivery trucks, or the people that
     were given primary consideration?) In what ways is this space justified
     economically? Is it possible or necessary to justify a non-commercial space
     in economic terms?

12   What is the basic shape of this space? Is there a dominant or organizing
     form? What are the auxillary forms used within it? Do you feel
     disoriented going through it? How would you alter the space to achieve a
     greater balance between order and variety?

13   What are the prevailing textures? materials? What are the basic colors?
     What seems to have been the designer's rationale for using these
     particular elements? Do the textures and colors complement and reinforce
     the forms? What color and texture do the people and their activities
     provide?

14   What are the large-scale patterns and rhythms in the space? Are the
     patterns created with forms, color, texture? Are they created by the people
     using the space? Are its elements in scale with you?

15   Do all these visual components work together in such a way that they
     can be grasped as a whole? Are they pleasurable to you as you walk
     through the space, as you run, ride a bicycle, drive?

Public activities require many
kinds of public spaces. *From left to
right:* pausing to enjoy the
splashing water of an urban
garden, Fort Worth; attending a
party in the Great Hall of the
Apparel Mart, Dallas; shopping
indoors in the Gallery,
Philadelphia; and strolling in the
arcade, Turin.

time logs assigned in Chapter 18, Personal Needs

a copy of "public spaces questionnaire" for each person

Participants should select a public space having both natural and built components and located within a convenient distance for a field trip.

further exploration

"Communicate in any medium how the people and their activities are used as components in the visual design of a chosen public space. Consider, for example, the patterns, color, and form created by clusters of activities and movement through the space.

"In the Milgrim reference, what did the authors find was the basis for recognizing locations within the city? What makes something memorable? In what ways was the field trip site recognizable within the city structure?"

check list

Was the participant able to analyze the composition of public spaces and their effect on him?

Was the participant able to mentally manipulate the components of built environments to improve the functional and psychological impact of public space?

Was the participant able successfully to communicate public space in his project?

resources

Bloom, Janet: "New Concepts for Public Space Combine Art and Architecture," *Architectural Record* (Feb. 1972) 101-140. Includes the work of SITE, Inc., who have developed brick and concrete buildings that are both sculpture and functional buildings.

Halprin, Lawrence: *New York, New York*, Chapman Press (1968). Analysis of the quality, character, and meaning of open space in urban design. Suggests ways to order the city so as to satisfy human needs.

Milgram, Stanley, *et al.*: "A Psychological Map of New York City," *American Scientist* (March-April 1972) 194-200. New cartographic techniques are proposed to map the psychological dimensions of cities.

Newman, Oscar: *Defensible Space*, MacMillan, New York (1972). How to design public spaces so they are safe.

Built and natural environment

25

Land development: a field trip to simulate the development of a natural environment into a built one reveals possible alternatives in working with nature.

we are part of nature

We tend to behave as if we are separate from, and in control of, nature, but we are not. Our effect on nature by our mere presence as living organisms must be considered in constructing any conceptual model of the environment.

Change in nature is continuous, with or without us. Nature destroys and builds. Our changes also can have extensive and long-lasting effects. But as far as we know, only man can predict the probable consequences of his actions (science) and can make value judgments about the effect of his acts (ethics). When or where to make a change in the environment is an ethical and moral decision, as well as a scientific one.

Any change in one environmental system causes change in the total environment. Housing built in the flood plain can act as a dam and cause the washout of buildings, creek banks, and bridges upstream. We too often ignore both scientific and ethical issues and act impulsively to solve short-range problems, such as how to get a quick return on an investment. Even when we attempt to make rational decisions, the complexity of nature ensures that changes probably will be produced by variables that we assumed were insignificant or which we did not perceive as part of the problem. Also, time may reveal an unexpected effect of a given change. A concrete channel can be built to control flooding, but this decision may overlook the ensuing increase in mass and velocity of water flow. In time, the faster stream flow resulting from the channeling may cause extensive erosion downstream.

You do different things in different climates on different land. In Albuquerque the typical planning solution *(left)* scratches off the mesa and remodels nature with planted trees and watered yards.

La Luz *(right and opposite)* is designed to protect the inhabitants from the negative aspects of the New Mexico environment, the harsh winds, the sun, and extreme dryness, and to protect the land from the people.

To conserve precious water and energy, the design and development work with nature. Houses are adobe and landscaping is sheltered in small walled courtyards.

inventorying a development site

Using the analytical processes developed during the previous study of a natural site, the participants should analyze this site before planning its development. They should include an inventory of existing characteristics *and of its surroundings.* Remind them to consider access to and egress from the site, the interface of the site with other properties, how the adjacent land is used, and the location of major drainage systems.

role-playing for planning strategies

"Remember that intelligent planning involves prediction. In working out your solution to developing this site, you will have to make value judgments about the probable effects of your changes. Take notes as you make your observations. You must be able to support your arguments before the council."

Divide the group into the three teams that will play roles according to the following instructions:
"Team I will plan for this site a project of any type they wish. As developers, however, they must keep in mind that they must plan a profitable venture to stay in business. The land may not be left unbuilt.

"Team II will act as a citizens' conservation group. The group may choose to subdivide into special interest groups such as the Audubon Society, sportsmen, and ecologists. Team II should take their inventory of information and establish priorities for preserving the natural environment. They should set criteria for development and plan a strategy to support, modify, or oppose Team I's project before the council.

"Team III will act as the city council, who must approve the developers' project before it can be built. They should, as a group, prepare to analyze the proposed project in terms of what it will both contribute and cost the city in money, human safety, and quality of life."

Ask the council members to convene and begin the hearing. First the developers present their proposal, then the conservationists deliver their reservations. The council must approve or disapprove the project with a full explanation of their reasons.

analyzing the decision-making process

The instructor may stimulate the discussion after the role-playing by interjecting questions such as these:

"Nature is too complex to work with in its entirety. Therefore to deal with the problem at hand, you must employ some selective process. Did you make a simplified conceptual model of the site and select significant variables from that model? If not, examine and explain what process you did use to identify all the variables you felt pertained. Could you make your case more easily if you reduced the number of variables? Have all three of the teams used the same variables in propounding their arguments for how the land should be developed?" Ask each team to prepare a list of the variables they feel are relevant. Rank the variables in order of importance. Are the three teams arguing from the same hierarchy of values? Compare the lists and discuss the differences. Did the vested interests inherent in the role a participant played influence which variables he considered significant when constructing his model? Can you as a group propose a development that would be an acceptable compromise to all three special interest teams?

"Should this land be developed at all? If not, does it fit into the city's over-all framework for land that should be preserved in its natural state? How would you categorize the 'natural state' of this land? Is it a true wilderness? Has it ever been forested, cultivated, or built upon? What evidence do you see to support your opinion? What economic mechanisms would you propose for keeping this land undeveloped?"

Summarize the discussion by asking "What are the important issues to be considered in any development? How are you going to work this information into your long-term project?"

preparation

Choose an outdoor environment with a variety of terrain and natural elements including a stream. An area with a minimum of alteration by man is preferable. The impact of this experience is heightened if the chosen site undergoes actual development during the period of this course.

further exploration

This learning experience can be extended to predict necessary resources such as water and energy for the inhabitants of the developed site.

Planning the development of the site may be expanded into a detailed, long-term exercise for groups such as architecture students.

check list

Is the participant consciously aware that as a human being, alive or dead, he is constantly and continuously changing the environment?

Has the participant understood that one must consider the *total* environment in planning any changes in a part of it?

resources

Bakeless, John: *The Eyes of Discovery*, Dover, New York (1950). Vivid accounts by the first white men to see North America.

Bedichek, Roy: *Adventures With a Texas Naturalist*, Doubleday, New York (1950).

Critchfield, H. J.: *General Climatology*, Prentice-Hall, Englewood Cliffs, N.J. (1966). Includes a discussion on the effects of weathering.

Eckbo, Garrett: *Urban Landscape Design*, McGraw-Hill, New York (1964). A view of landscape as the result of interaction between man and nature.

The Fitness of Man's Environment, Smithsonian Institution Press, Washington, D.C. (1967). Papers presented by biologists, architects, anthropologists, and other professionals at a symposium held to seek guidelines for comprehensive solutions to environmental changes.

McHarg, Ian L.: *Design With Nature*, American Museum of Natural History, New York (1969). The interaction of the built with the natural environment; problems and possible solutions.

Mesarovic, M., and Pestel, E.: *Mankind at the Turning Point*, Dutton, New York (1974). This second annual report to the Club of Rome is appropriate reading for a systems perspective on nature and man.

Open Space, Identifying the Critical Areas, prepared by the North Central Texas Council of Governments, Arlington, Tex. 76011 (1975). An outstanding resource to help lay people learn the terminology and techniques that will enable them to evaluate the natural features of land and to participate in the planning process.

Ward, B., and Dubos, R.: *Only One Earth*, Norton, New York (1972). An unofficial report by the United Nations Commission on Human Resources that emphasizes the need to conserve our natural resources for the generations to come.

Dealing with change

26

Orienting in the context of change: the group examines the meaning of change for the individual and society.

change is here to stay

Change is inevitably with us. Changes in the world around us can be viewed as neutral, not of themselves good or bad; they happen. The individual makes a personal judgment about a change. The judgment depends on how he perceives the change operating in a particular context and how he projects its consequences as being a personal gain or loss.

Both the magnitude of the change and the rate of change determine our perception of a change. A change can be as trivial as a new pair of shoes or momentous as the burning of one's home. Our reaction can range from ecstasy to despair. If the rate of change is slow enough, its variables can be understood, the new milieu becomes predictable, and one can orient oneself to the change. When the rate of change becomes too great, when transition without apparent beginning or end is all we perceive, orientation cannot easily take place and we are left with anxious and ambivalent feelings. Eventually we will adapt to the change. We can choose to orient ourselves to the new set of circumstances: we can make the change part of our world by accepting it or by taking action to modify or redirect it.

Everyone adapts to change but not everyone is able to orient. The process of orienting includes knowing who and where we are at a given moment in relation to our social and physical world. Change is with us constantly. We all develop patterns for responding to change. An individual's personality — particularly his flexibility, his age, and his culture — influences the development of these patterns. The individual who understands his patterns of dealing with change has a better chance to orient. Rather than resisting change, he opens up new possibilities for enhancing the quality of his personal life.

The rate of change in our society was never so great; the magnitude of change seems more and more overwhelming. Day before yesterday we were at war, yesterday we had an environmental crisis, today we have an energy crisis, tomorrow . . . We have choices to make.

change in your life

"What does not change? Try to think of ten things that do not change. Analyze carefully each suggested item. Does it really not change at all, or is this an instance of no apparent or significant change?

"How would you feel if there were no change at all? How do we benefit from change? Is part of the value of change the relief from boredom? What are your favorite changes? What changes do you look forward to?

"We have repeatedly looked at change related to the environment. Change occurs in ourselves from life to death, continuously. Those changes in ourselves influence how we perceive and judge changes in our surroundings.

"In what ways have you changed in the past year? What do you do differently? What attitudes and ideas have changed?" (Sometimes a change in attitude is easier to perceive over a five-year period.)

"How many of you in the past six months or year have had a change in residence, car, job, hair length, father's job, girlfriend, spouse, weight, eating habits? How many feel *you* have changed in this same period of time?

"What has been the most momentous change for you in the past year? How did you react to this change? Did you view it as 'good' or 'bad' then? How do you view it now? Was this change the result of your action, or was it decided for you?

ca. 1873 for 8 years

1889 for 25 years

"Are you afraid of change? Why?" (You may decide in advance that the change will be 'bad'; for example, that a loved one will be dead. Or you may feel that the change will cause you to lose some control over your own life. For example, if you suddenly can't afford to drive a car, you feel you can't go places you want or need to go.) "Has an impending change ever caused you anxiety or uneasiness because you didn't know whether the effect would be 'good' or 'bad'? What makes you anxious?" (If the rate of change becomes too great, we lose the ability to orient with the change.) "Do you ever resist change? Why?" (You may have a vested interest in the way things are simply because you know the consequences of the present.) "Have you ever forced yourself to make a change, to take the implied risk?

"You may feel anxious or ambivalent about a change. By orienting to this change — that is, by making it a part of your life — you can lessen your internal conflict. The better you understand your own patterns of responding to change, the easier the orientation will be.

"How do you deal with change? Do you try to predict extremes that are the best and the worst that can happen? What kind of change brings forth these responses:

'If only . . .' (attempt to reverse the change)
'It makes me sick . . .'
'Hooray!'
'Do you really mean it?'
'I don't believe it.'
'It's about time!' (a delayed change)
'I'll get even with you!'
Frustration (a feeling of being helpless to deal with the change)
Extreme anger (usually leads to despair or denial)

"Think of someone you have observed deal with change. How does he or she do it? (For example, if one is able to fit change into an existing pattern, one's anxiety greatly decreases. The fact that changes in nature occur in a rhythmic pattern makes them predictable. Thus they are easier to handle, even though one is not always prepared for an extreme fluctuation in the pattern, such as a hurricane.)

"What changes in your life do you feel you dealt with successfully? Select an important and recent change; analyze it in detail. What enabled you to deal successfully with that change? What existing patterns did you use? Did you have to alter your old patterns at all?"

1914 for 42 years

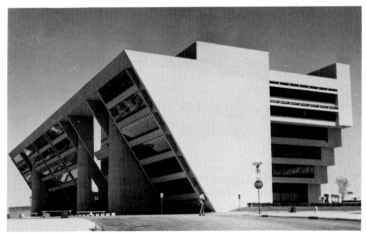

1978 for ?

change in a historical context

Ask the participants to begin observing change from a sociological and historical perspective.
"In our society, change, because it represents progress, has traditionally been valued. The opposite was true of conservative societies such as ancient Egypt. The rate of change in our society during the last hundred years is phenomenal. Our society has chosen to stimulate growth and change. How have we benefited from change? What kinds of difficulties have resulted from our valuing change?

"Return to the neighborhood visited and interview its long-term residents. Try to collect data that will shed light both on the physical changes that have taken place in the last ten to twenty years, and the residents' attitudes toward changes that have taken place or that may one day take place."

The Dallas city government, after renting space for its first seventeen years, built quarters over a meat market in about 1873. In 1884 the city offices moved to space housing the fire department. Only in 1889 did Dallas finally have its first unshared city hall. In 1914, 1956, and again in 1978 expansion of city services prompted rebuilding. How soon will this latest monumental structure be replaced?

revisiting the neighborhood

"Before returning to the neighborhood to interview the residents, agree upon the questions you will ask. Your questionnaire might include, for example:

1. What changes in buildings have taken place in this neighborhood in the last ten to twenty years?
2. What changes in nature have occurred? Were the changes caused by people? What changes in nature have had an effect on the people living in the neighborhood?
3. What changes do you think will come to this neighborhood in the next ten to twenty years? What changes would you like to see take place?
4. If you had a choice, in what fifty- or one-hundred-year period of the past, present, or future would you prefer to live? Why? (The answer to *why* gives as many clues to what an individual personally values as his original answer to *when*.)
5. Do you think our country will be a better place to live in thirty years? Why? Why not? (Note the reasons given for the answers. Their explanations will reveal the respondents' attitude toward change.)"

With the reassembled group, evaluate and discuss the raw data collected. During the discussion be certain to analyze the limitations of the data. If, for example, only a few people were interviewed and they were all from the same location and of about the same age, the data will be biased.

what will we be?

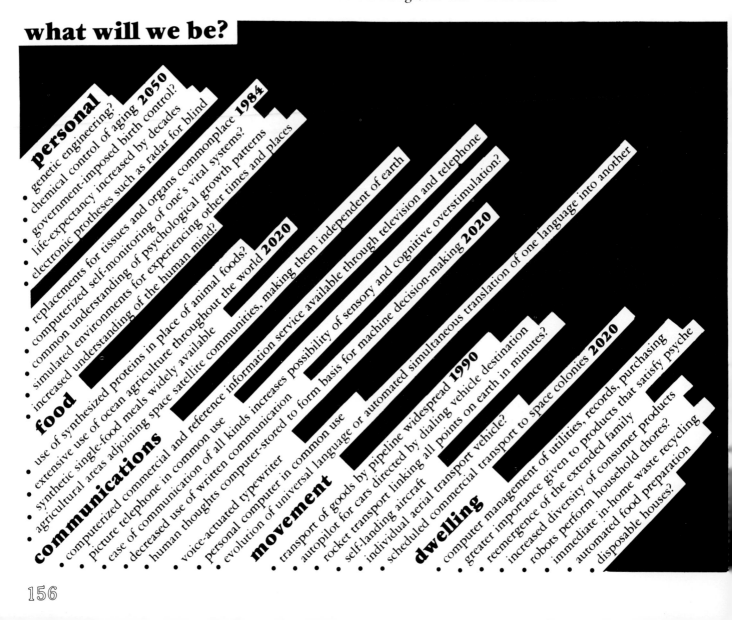

personal
- genetic engineering?
- chemical control of aging **2050**
- government-imposed birth control?
- life-expectancy increased by decades
- electronic prostheses such as radar for blind
- replacements for tissues and organs commonplace **1984**
- computerized self-monitoring of one's vital systems?
- common understanding of psychological growth patterns
- simulated environments for experiencing other times and places
- increased understanding of the human mind?

food
- use of synthesized proteins in place of animal foods? **2020**
- extensive use of ocean agriculture throughout the world
- synthetic single-food meals widely available
- agricultural areas adjoining space satellite communities

communications
- computerized commercial and reference information service available through television and telephone
- picture telephone in common use
- ease of communication of all kinds increases possibility of sensory and cognitive overstimulation?
- decreased use of written communication **2020**
- human thoughts computer-stored to form basis for machine decision-making
- voice-actuated typewriter
- personal computer in common use
- evolution of universal language or automated simultaneous translation of one language into another

movement
- transport of goods by pipeline widespread **1990**
- autopilot for cars directed by dialing vehicle destination
- rocket transport linking all points on earth in minutes?
- self-landing aircraft
- individual aerial transport vehicle?
- scheduled commercial transport to space colonies **2020**

dwelling
- computer management of utilities, records, purchasing
- greater importance given to products that satisfy psyche
- reemergence of the extended family
- increased diversity of consumer products
- robots perform household chores?
- immediate in-home waste recycling
- automated food preparation
- disposable houses?

reexamining the city

During the field trip in the session on Orienting in the City, participants were asked to note on a map major changes in appearance and in land-use patterns that had taken place in the last ten to twenty years. Use the information recorded on the maps as raw data. If the data are incomplete, it may be desirable to return to the site. Many people are inexperienced in this type of data collecting and do not "see" what they are looking at during the first visit.

"Use the data to help answer questions about change in your city. What are the principal changes that have taken place in the past ten to twenty years? What social attitudes, technological innovations, and economic factors help explain these changes? What indirect controls influenced the location of freeways?

"What changes in the physical landscape do you foresee?" (the use of mass transit, more separation of light and heavy industry, disappearance of parking lots . . .)

what is the future bringing?

"We live in transition from one state to the next. It is difficult to see ourselves living as we will in the future. Use the two charts as guides to make a list of changes you would like to see. Now add changes that may evolve that will cause more problems than they solve. What kind of people will we become as a result of growing up and living in this projected future?"

charting your future

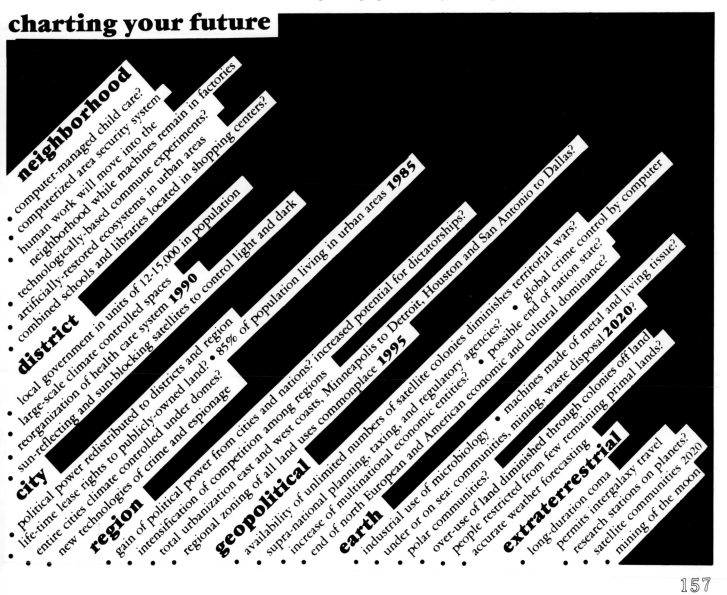

neighborhood
- computer-managed child care?
- computerized area security system
- human work will move into the neighborhood while machines remain in factories
- technologically-based commune experiments?
- artificially-restored ecosystems in urban areas
- combined schools and libraries located in shopping centers?

district
- local government in units of 12-15,000 in population
- large-scale climate controlled spaces **1990**
- reorganization of health care system
- sun-reflecting and sun-blocking satellites to control light and dark

city
- political power redistributed to districts and region
- life-time lease rights to publicly-owned land? • 85% of population living in urban areas **1985**
- entire cities climate controlled under domes?
- new technologies of crime and espionage

region
- gain of political power from cities and nations? increased potential for dictatorships?
- intensification of competition among regions, Minneapolis to Detroit, Houston and San Antonio to Dallas?
- total urbanization east and west coasts, Megalopolis commonplace **1995**
- regional zoning of all land uses

geopolitical
- availability of unlimited numbers of satellite colonies diminishes territorial wars?
- supra-national planning, taxing, and regulatory agencies? • global crime control by computer
- increase of multinational economic entities? • possible end of nation state?
- end of north European and American economic and cultural dominance?

earth
- industrial use of microbiology • machines made of metal and living tissue?
- under or on sea: communities, mining, waste disposal **2020**?
- polar communities?
- over-use of land diminished through colonies off land
- people restricted from few remaining primal lands?
- accurate weather forecasting

extraterrestrial
- long-duration coma permits intergalaxy travel
- research stations on planets?
- satellite communities 2020
- mining of the moon

157

preparation

Arrange to return to the neighborhood previously visited.

maps used in Chapter 22 during the field trip "bird's eye viewing"

further exploration

"Write an essay on changes you would like to see in your immediate environment. Analyze how these changes might be made."

Ask the participants to repeat the Chapter 14 photostudy field trip. Give them the same worksheet, but ask that the questions be applied specifically to the urban environment. (The results of the exercise carried out at two different times can be compared to evaluate the participants' learning during the course.)

check list

Can the participant articulate the patterns he uses in adapting and orienting to change?

Can the individual suggest processes that will help him orient more readily to the increased rate of change associated with global living?

resources

Blackmore, John: "Community Trusts Offer a Hopeful Way Back to the Land," *Smithsonian* (June 1978) 97-109. A social development that encourages new forms of land ownership to meet both economic and social needs.

Fabun, Dan: *Dynamics of Change*, Prentice-Hall, Englewood Cliffs, N.J. (1967).

Future Shock, film produced by Metromedia Producers Corp. (1972). Distributed by McGraw-Hill. Award-winning film narrated by Orson Welles.

Helmreich, Robert: "Evaluation of Environments: Behavioral Observations in an Undersea Habitat," *Designing for Human Behavior*, J. Lang et al. (editors), Dowden, Hutchinson and Ross, Stroudsburg, Pa. (1974). A description of the concepts, technology, and methodology used to make behavioral observations.

Lynch, Kevin: *What Time Is This Place?* The MIT Press, Cambridge, Mass. (1972). An examination of the human being amid change and the use of planning for the management of change.

Meadows, D., Meadows, D., Randers, J., and Behrens, W. III: *The Limits to Growth*, Universe Books, New York (1972). A report for the Club of Rome's project on the predicament of mankind.

Michener, James: *The Source*. Thousands of years of change evolve around the continuity provided by a source of water in this novel.

Müller, Jörg: *The Changing City*, Atheneum, New York (1977).
Müller, Jörg: *The Changing Countryside*, Atheneum, New York (1973). Two portfolios document in detailed, fold-out pictures, the changing scene in a typical city and in a countryside.

O'Neill, Gerard K.: *The High Frontier: Human Colonies in Space*, Bantam Books, New York (1978). A projection of what is technically possible now and for the future.

Swann, Robert: *The Community Land Trust: A Guide to a New Model for Land Tenure in America*, Institute for Community Economics, Cambridge, Mass. (1972). A more extensive treatment of this idea than is given in the *Smithsonian* article cited above.

Toffler, Alvin: *Future Shock*, Random House, New York (1970). A convincing argument that we will have to learn to adapt to greatly accelerating rates of change.

Taking responsibility for environment

27

Analyzing components within the whole: participants summarize their experiences, applying the ideas, points of view, and information dealt with in the course.

space ship earth is finite?

Historically, we have overused the land and then moved on. We have used the resources of our planet as if they were limitless. We have sought physical frontiers to conquer, then consumed them with abandon. Each decade we "discover" that the information we have been using to guide our consumption has been at best incomplete, at worst wrong. We cannot say assuredly, "Tomorrow's knowledge will solve our problems." Nor can we say, "If we continue our consuming ways, we are doomed." We do not know. We do know there will be changes — changes in our knowledge, technology, perceptions, habits, values, and behavior.

Responsibility cannot be turned over to "them" or a government. Responsibility must be immediate and personal, for it is the sum of our acts that downgrades our planet as an environment for living. We must consider our actions on all scales, from minor ingrained habits to major considered policy decisions. What price are we paying for the pleasant cascade of water that accompanies the brushing of our teeth? To what future trade-offs are we committing ourselves as we permit building in our flood plains?

Environmental issues are social, and therefore political. We act with passion when our personal property is threatened by a new freeway. Only a high level of environmental awareness can foster a similar response to the routing of the same highway through someone else's neighborhood. When we are able to understand our personal stake in environmental changes that are seemingly remote, we may be motivated to commit time and energy to the city as a whole. To act effectively we must assess proposed changes and consider alternative solutions. Otherwise, issues remain merely irritants in the newspaper.

First we must see clearly what the issues are. An awareness of the environment enables us to consider more variables and thus to examine each problem more broadly. The greater our understanding of how things interrelate, the better our chance of seeing alternative solutions. As we analyze environmental problems in increasingly sophisticated ways, we conceive more effective courses of action.

Problems of change as we perceive them must be communicated to others to be effectively solved. Learning to communicate decisions and the rationale they are based on is a part of decision-making. Learning to communicate is critical to our active participation in society — the context of all environmental issues. Verbal communication about the environment, because it deals in symbols (words), is inadequate to convey the total experience. Visual communication combined with verbal is still at best a "two-dimensional" communication. As such it cannot convey how people experience the environment or how it changes with time. Each of us must discover his personal method of transmitting as much as possible of the whole situation.

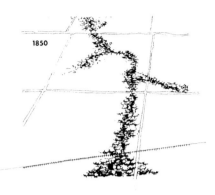

1850

In taking responsibility for our environment each of us must choose his role. Will it be personal conservation — saving bottles and cans? Will it be professional commitment as a geographer, a developer, urban planner, or biologist? Will it be interaction with a group on specific community issues — flood plain management, zoning codes . . .?

When by choice we do get involved, to arrive at what we judge — on the basis of incomplete data — to be the best workable solution, we are forced to weigh conflicting goals. Where will we set the line of compromise, for example, between the need for affordable energy and the need for an unscarred, unpolluted land? It is up to us to decide.

strategy for action

"Each of you choose the environmental issue that concerns you most directly and personally. 'Energy' is too broad an issue; a proposed body shop adjacent to your home would be more appropriate. Other ideas to explore might be a freeway proposed through 'your' park, a zoning change in your neighborhood, or a proposed dam on a river down which you canoe. Choose an issue you care about."

Above, **1840's.** Indians travelled along this creek to a river crossing *upper left of this view.* The boundaries of land grants gradually became roads as the land was fenced. Water in the Trinity Sands (3200 feet below surface) gushed 50 feet into the air when tapped. Settlers did not need cattle for meat in their first years because of the abundant wildlife.

1850's to 1940's. The land was farmed. A railroad was built in the 1880's *(diagonal line at bottom).*

1940's to 1960's. Land was bought for speculation, in anticipation of urbanization, and used for recreation (horse farms and weekend retreats).

Right, **1973.** The two-lane roads have been replaced by a freeway, *left,* and a six-lane thoroughfare, *right.* A warehousing district with rail connections is well along after five years of development. The 1880's railroad is now a railhead. Where once it continued, a toll road has supplanted its right-of-way into the heart of the city *(diagonal line at bottom and out of picture to the lower left).*

Car showrooms and retailing begin to line the freeway, *lower left.* A Treasury and steak house begin construction on Midway, *left center.* Housing has backed up to the freeway, *upper left.* A pedestrian bridge that spans the freeway, and sidewalks within some neighborhoods, are the only corridors for pedestrians.

1973

Apartments straddle the creek, *upper center.* The old grid has been interrupted by curving streets through residential areas.
The prep school that moved to the country in the 1950's, *corner of Midway and Spring Valley, right center,* now adjoins a warehouse district *below the school,* and a public school field house *above it.* The

creek is still visible, but it dies abruptly at the school property; it was filled and channeled this side of Midway Road when the industrial area was developed beginning in the late 60's.

1975. A community college has forced the City of Farmers Branch to relocate Alpha Road planned to

cross the college property along the creek. Single family homes are built one foot above the hundred year flood plain line, *just to the right of the college.*
1976. The community college begins construction. Water in the Trinity Sands that formerly gushed when tapped, now rises only to 500 feet below the surface.

"Document the environmental problem you have chosen. State in detail your solution to the problem. List persons and organizations you think would support your position. List those who would favor other solutions and their reasons for doing so. What compromises can you see among the various solutions? Realistically, what could you do if you so desired? Write out a plan of action detailed enough to include names of friends, organizations, councilpersons, and others you could enlist in your cause. Itemize steps you would take. Discuss and compare your strategy with others."

presentation of projects

Ask the participants to present their long-term projects. Each individual or team should introduce their project with their original statement of purpose. The rest of the participants should analyze the extent to which each project achieves its stated purpose. "Which was the most commonly chosen target audience for your projects? Which communication technique was chosen most often? Which technique communicated the most clearly? Did any project address more than the visual and aural senses?" The participants should examine in what ways each project met its own goals and in what ways it did not. "What suggestions do you have for helping each project communicate?"

The college for 4000 students opens. Alpha Road is extended and turned to follow the college boundary. The nearest public transit is two miles.

1980's. Prognostication: Highrise office buildings will replace low retail fronting the freeway. The impacted intersection will be rebuilt. The college will expand to 10,000, later to 20,000 students. Cheaply built apartments will begin to decay. Thoroughfare volumes will triple. Pressure to add frontage roads and commercial uses will downgrade residential districts adjacent to the freeway, *upper left.*

Stormwater runoff from the warehouse area, unable to percolate into the ground through parking lots, streets, and buildings, will increase downstream water speed, eroding stream banks and backing up at narrow places. The chances of flooding homes built low, next to the creek, will force straightening the creek (or moving the houses, usually politically unfeasible). Channeling the creek and reducing the vegetation cover will further limit wildlife; the food chain will be cut, removing all but hardy insects and birds. Reduced percolation of water into the ground will dry up the springs and lower the water table.

These are the results of our past decisions. How will we organize for change in the future?

1977. Two coyotes, rabbits, and wildflowers sighted on college site, *upper left.* Springs still flow from the creek banks. Runoff is being increased in the creek by 100% from the warehousing and other development. Municipal boundaries of three communities complicate coordination of land uses. The City of Addison has zoned for apartments; the City of Farmers Branch, for warehouses separated from its own single family housing. Farmers Branch has built Midway Road into a heavy trucking thoroughfare, but Addison has not cooperated in extending it, *right.*
A first highrise is built, *lower left,* then a second, *center left.*

1978. A third highrise is built, *center left.* The owner of the vacant land *above the college site* announces development of single family homes. The remaining undeveloped quarter section is held by a major oil company. Traffic consistently backs up 800 to 1000 feet at Midway and the freeway intersection, *left center.*

other options

We have the option to move to new frontiers. Will we move on or under the sea, to Anartica or to satellite colonies in space? As we have learned to control more and more factors in our own environment, we have become better equipped to build totally artificial environments. What if we treat space not as a void, but as a "culture medium rich in matter and energy"?

Gerard O'Neill at Princeton University has calculated that a pilot satellite colony is capable of being built now. The technology is available. The O'Neill group proposes colonies in space with energy supplied by the sun, raw materials by the moon and asteroids, and with gravity simulated by rotational acceleration. O'Neill's conception envisions technologically and socially self-sufficient governmental units with cultural diversity and a high degree of independence from earth. A community of 200,000 people desiring to preserve its own culture and language could choose to remain isolated.

Apparently the environment of earth including mountains, clouds, and animals, could be recreated and possibly even improved. The need for pesticides would be eliminated because of protected farming areas and selected seeds.

What would life in a space colony be like? Psychologists feel it will be necessary to offset negative feelings of crowding and isolation from earth. Will such qualities as color, texture, and light assume even greater importance in achieving "positive arousal to the environment"?

Psychological testing on earth suggests that isolated groups need to be able to alter parts of their environment at will. A sense of isolation can also be decreased if there is maximum visual access to an outside environment, whether it is under the sea, or in space. In a world totally built by ourselves will we find enough that is unknown and unpredictable to keep us involved and interested?

Will we want to recreate the environment of earth? Will we choose to repeat our twenty-four hour and three-meal day? The precedent of the Eskimos tells us

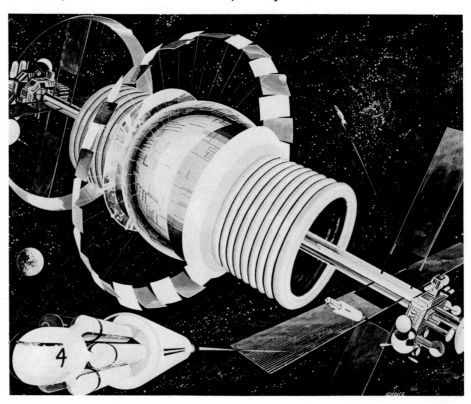

From the exterior, NASA's conception of a space habitat is seen to be a sphere nearly a mile in circumference that rotates to provide a simulated gravity at the "equator" comparable to that on earth. Docking areas and zero gravity agricultural and industrial areas are at each end of the space community. The flat surfaces radiate away waste heat.

that this is neither a biological necessity nor universally adopted on earth. Or will we choose the anthropologists' vision of an opportunity to try alternative ways of setting up new frontier communities? We may choose a different structure of time, of physical space, of political and social units. We could build a large number of small — or a small number of large — communities.

By the end of the century, payload delivery costs are expected by some scientists to be low enough to make space manufacturing facilities feasible. Will we human beings be ready to pioneer space that soon? We will have old problems to solve in new locations amplified by increased tensions caused by isolation. Will we finally be able to answer the question whether the "right" environment can solve such problems as juvenile delinquency, poverty, and crime? How will we determine what that "right" environment might be?

In the past, wars have been fought to acquire the most desirable land. Will there be new forms of desirable land in space? Or does the unlimited territory in space imply high quality of living for all peoples with no further need for territorial wars? Will warfare, nevertheless, prove to be an inherent part of our lives?

If we immigrate to distant colonies we will also lessen the pressures upon earth. Will the opportunities the space colonies give us eliminate the problems we perceive, such as overpopulation, limited natural resources, and conflicting ideologies? Can we transfer our drive for growth to space colonization, leaving us free to conserve our resources on earth?

Will we change? Our past experience with Utopias, communes, or other isolated communities gives us no reassurance that a change of location will change *us*. We have opportunities to shape our environments in these new situations and therefore shape ourselves in ways undreamed. As we encounter new environments, and reencounter the environment on earth, what decisions will we make?

The interior of the 10,000 person settlement intended for space manufacturing appears very much like our environment on earth.

For educators

Environmental Encounter is a vehicle to help people learn to make informed decisions about their environment. The over-all strategy is to lead the participants through a series of experiences so exaggerated that they will be able to organize and cognify their impressions. The experiences are so designed that the impact the individual has on the environment as well as the impact the environment has on the individual is apparent. Throughout the book the participant's personal experience is turned into the experience through which the teaching and learning take place.

The content in the twenty-seven chapters is sequenced so that the concepts learned in the earlier chapters are applied to increasingly complex urban environments. As material is grasped, it becomes part of the participants' experience, enabling them to consider environmental issues in greater scope and detail.

learning experiences

The teaching core of each chapter is a group of *learning experiences*, designed to make visible the concepts discussed in the chapter introduction. The learning experiences can introduce or accompany lecture topics, design problems, or readings as part of your course syllabus. The exercises can be directed by you, led by teaching assistants, or in many cases, carried out by the participants themselves acting as self-instructors.

The learning experiences can also be used to meet the more limited goals of workshops or seminars. Adults can complete many of the exercises in fifteen minutes to an hour; younger people usually require more time. These learning experiences may suggest modifications or supplemental learning strategies for your particular course.

Because the experiences are direct interactions of the participant with his environment, they are appropriate for individuals of diverse backgrounds. Discussion that follows the exercises can be specifically directed to meet your goals and to fit the background and interests of your students.

analysis and discussion

The analysis and discussion after each experience provide two opportunities. They allow the participants to draw on their past experiences in developing fuller understanding: they become active agents in their own learning process. They also give the instructor an opportunity to direct the learning according to *his* expertise. The educator helps the students extrapolate from their classroom experience and apply what is learned to real world problems. The educator sets the level of sophistication of the analysis. The environmental interaction takes place; it is guided by the instructor's expertise; the participant integrates new insights with past experience.

teaching notes

Marginal comments called *our experience* summarize teaching techniques that worked well in pilot tests with particular exercises. *Preparation*, annotated *resources*, and suggestions for *further exploration* are included at the end of each chapter for your convenience.

evaluation

For evaluating the teaching and learning experience, two chapters — Photointerpretation and Sequence of Spaces — and a suggested long-term project have been included. Some course leaders have found it helpful to require journals. Others have used the questions titled *check list* following each chapter. Other evaluating methods to meet more specific goals can be developed.

The course provides a framework in which participants can organize new information acquired from any source. It encourages students to take responsibility for their learning, their decisions, their actions, and their environment.

illustration provenance

index